A User Guide on

PROCESS INTEGRATION FOR THE EFFICIENT USE OF ENERGY

Revised First Edition

Authors of the Guide

B. Linnhoff	UMIST
D.W. Townsend	ICI New Science Group
D. Boland	ICI Petrochemicals and Plastics Division
G.F. Hewitt	AERE Harwell
B.E.A. Thomas	Process Engineering Consultant
A.R. Guy	Filtration and Transfer Ltd
R.H. Marsland	Johnson-Hunt Ltd

Contributors

J.R. Flower	University of Leeds
J.C. Hill	ICI Petrochemicals and Plastics Division
J.A. Turner	ICI New Science Group
D.A. Reay	International Research and Development Co Ltd

The authors were members of a Working Party set up by the Institution of Chemical Engineers to prepare this Guide under the Chairmanship of B.E.A. Thomas.

INSTITUTION OF CHEMICAL ENGINEERS

Published by
Institution of Chemical Engineers
Davis Building
165–189 Railway Terrace
Rugby, Warwickshire CV21 3HQ, UK.

© Institution of Chemical Engineers 1982, 1994
A Registered Charity
© The diagrams in Sections 1.1 and 2.2 are the copyright of
Professor B. Linnhoff (originally ICI)

Reprinted 1994

ISBN 0 85295 343 7

Printed in the United Kingdom by Warwick Printing Company Ltd, Theatre Street,
Warwick CV34 4DR.

Contents

Contents

Foreword to the Revised First Edition

Twelve years have passed since the *User Guide on Process Integration* first appeared. Since then, the world has moved on. Contrary to expectations, the cost of energy (in real terms) has come down. Nevertheless, the concept of 'efficiency' in process design has grown more important. This is mainly due to two trends: the industry's response to environmental issues, and a general drive for quality in the broadest sense. Both trends have touched the life of the professional in all corners of the industry, including process design.

Against this background, process integration has evolved, and its application has evolved. Twelve years ago, it was a specialised tool. Today, it is applied routinely. It has become 'good practice'. It has become integral to the quality design process aimed at lower emissions, reduced investment, etc. Its scope is broader and it no longer represents a specialism.

Throughout this evolution, the original concepts of Pinch Technology have stood the test of time. Composite Curves, Grand Composite Curves, the Grid Diagram, the Pinch Design Method, Appropriate Placement, etc. are still used today as they were used twelve years ago. But there are now additional concepts. The new concepts have broadened the technology from a heat recovery tool into a general process analysis and design tool. There are 'Shaft Work Targets', there are distillation 'Column Profiles', there are 'Pressure Drop Targets', there is 'Total Site Integration', there is '(Waste) Water Pinch', there are 'Emissions Targets', there are concepts for batch process design, for design for multiple base cases, etc.

This 1994 Revised Edition of the original *User Guide* has not been revised as far as the original concepts are concerned. There is however the addition of a recent review article originally published in the IChemE's *Transactions, Part A (Chemical Engineering Research and Design)*. This article gives a brief overview of the original concepts covered in the main text of the Guide, followed by an outline discussion of the recent developments mentioned above, of industrial applications, and of alternative approaches to process integration (based on mathematical programming, etc.). The review article is independent from the main body of the book — it can be read before, during or after. May it whet your appetite for professional fun in the design of quality processes and total sites!

Professor B. Linnhoff
Head of Department
Department of Process Integration,
UMIST, Manchester

Foreword to the First Edition

Every now and then there emerges an approach to technology which is brilliant — in concept and in execution. Of course it turns out to be both simple and practical. Because of all these things it is a major contribution to the science and art of a profession and discipline.

Bodo Linnhoff and the other members of this team have made a major contribution to chemical engineering through their work. It is already recognised worldwide and I have personal experience of the acclaim the techniques embodied in this guide have received in the USA.

There is no need to underline the necessity for more efficient use of energy: the chemical industry is a very large consumer, as a fuel and as a feedstock. What is equally important is that conceptual thinking of a high order is necessary to our industry to keep advancing our technologies in order to reduce both capital and operating costs. The guide provides new tools to do this, which forces the sort of imaginative thinking that leads to major advances.

It is also important to note that the emphasis in the guide is on stimulating new concepts in process design which are easily and simply implemented with the aid of no more than a pocket calculator. In these days, when the teaching and practice of many applied sciences tend heavily toward mathematical theory and the need for sophisticated computer programs, a highly effective, simple tool which attains process design excellence is very timely.

R. Malpas
President and Chief Executive Officer
Halcon International Inc.

1. Introduction

Figure 1.1a shows an outline flowsheet representing the traditional design for the front end of a specialty chemicals process. Six heat transfer "units" (*i.e.* heaters, coolers and exchangers) are used and the energy requirements are 1722 units for heating and 654 units for cooling. Figure 1.1b shows an alternative design which was generated using the techniques for network integration introduced in this Guide (Linnhoff *et al*, 1979). The alternative flowsheet uses only four heat transfer "units" and the utility heating load is reduced by about 40% with cooling no longer required. The design is as safe and as operable as the traditional one. It is simply better.

This example demonstrates the way in which the techniques introduced in the present Guide improve designs. Improvements are not due to the use of advanced unit operation technology, but are due to the generation of simpler, more elegant, and more appropriate integration schemes.

There are two engineering design problems in chemical processes. The first is the problem of unit operation design and the second is the problem of designing total systems. The Guide addresses itself to the system problem.

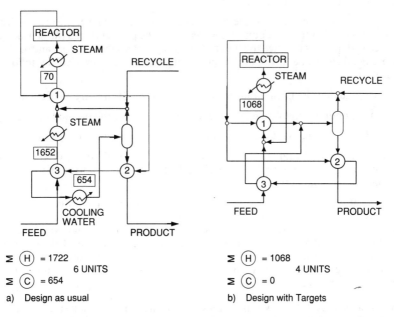

Figure 1.1—Outline flowsheet representing the traditional design for the front end of a specialty chemicals process

Past Record

The techniques for integrated network design presented in the Guide were developed over the past ten years or so at the ETH Zurich, Leeds University, ICI and UMIST (for example, Linnhoff, 1979; Linnhoff and Flower, 1978; Linnhoff and Hindmarsh, 1982; Townsend and Linnhoff, 1982a). Over the past six years, they have been tried extensively on live design projects in ICI and have become recognised in the industry as a development of great value and practicality. In a recent paper (Linnhoff and Turner, 1981) results from applications in ICI were published; see Table 1.1. This Table lists case studies performed between 1977 and 1981. It indicates the area of technology, whether the project in question was aimed at the modification of existing plant or the design of new plant, the available energy savings identified, and the capital cost implications of the relevant design changes.

Note that in at least half of the new design projects, capital cost implications were savings rather than expenditure: in other words, more elegant integration made it possible to identify processes that would not only be cheaper to run but also cheaper to build.

This is a central theme running through the Guide. We are all used to believing in the inevitability of an energy-capital trade-off: we retreat in one respect and advance in the other. However, we must realise that this notion rests on the assumption that there is no basic fault in our design to begin with. As we will see in the main text of the Guide, we might well save energy *and* capital if our starting point is non-optimal in both respects for no good reason!

Due to changes in ICI's corporate economic strategy only some of the projects listed in Table 1.1 have so far been built. However, in all the projects listed the new designs have been adopted as practical by the relevant design teams (contractors and/or ICI). A more detailed evaluation than that given in Table 1.1 has revealed the following picture:

● Energy savings identified ranged from 6% to 60% of the original design.

● Corresponding capital savings were as high as 30% of the original design.

● Payback times in plant modifications were improved by factors of up to four.

● The improved designs do not tend to present unusually difficult control and operability problems and indeed some featured better control characteristics than the original designs.

The applications have covered a wide variety of technologies (continuous and semi-batch, hydrocarbons and general chemical) and a wide variety of design objectives (new design and debottlenecking, optimisation and trouble-shooting). More recently, companies other than ICI have reported successful applications (Cooper, 1981). Many of these applications stem from the refining industry while others cover a cross-section of the chemical industry. In this Guide, case studies are discussed relating to the preheat train of a crude unit, an aromatics process and an evaporator/dryer plant.

Why this success? The two main reasons are that the techniques are based on completely new principles and concepts and that they are simple. The heat recovery

"Pinch" is a new thermodynamic concept at the centre of the techniques. Following on from the "Pinch" there are many related new concepts and techniques. Between them, these concepts and techniques make it possible for the user to deal quickly and confidently with problems hitherto thought too complex to be properly understood.

Conceptual Understanding and Simplicity

Systematic design of integrated process networks has been an active research area in universities for a few years now. It is worth pointing out that the techniques presented in this Guide have little in common with most previous techniques developed in this area. Most of the earlier techniques were aimed at press-the-button computer implementation and relied on powerful mathematics to examine a large number of alternatives generated by "rules of thumb". The techniques presented in this Guide place an emphasis on conceptual understanding and practical thermodynamics rather than rules and on simple sums rather than powerful mathematics. With this approach, the apparent complexity of the problem is reduced so drastically that the need for computers disappears. Pocket calculator programs might be convenient for some of the more repetitive calculations but are not essential. Trained users invariably find the techniques both mentally stimulating and quick and easy in application.

Table 1.1—Results of applying network analysis to projects

Process	Facility*	Energy savings available $/yr	Capital cost expenditure or saving $
Organic bulk chemical	New	800 000	Same
Specialty chemical	New	1 600 000	Saving
Crude unit	Mod	1 200 000	Saving
Inorganic bulk chemical	New	320 000	Saving
Specialty chemical	Mod	200 000	160 000
	New	200 000	Saving
General bulk chemical	New	2 600 000	Unclear
Inorganic bulk chemical	New	200 000 to 360 000	Unclear
Future plant	New	30 to 40%	30% saving
Specialty chemical	New	100 000	150 000
Unspecified	Mod	300 000	1 000 000
	New	300 000	Saving
General chemical	New	360 000	Unclear
Petrochemical	Mod	Phase I 1 200 000	600 000
		Phase II 1 200 000	1 200 000

*New means new plant; Mod means plant modification.

Structure of the Guide

In Section 2 a step-by-step description is given of the basic network integration principles both for process and utility networks incorporating either heat exchangers on their own or turbines, heat engines and heat pumps alongside heat exchangers. Much of this material follows in outline a training course which is on offer regularly to industrial design engineers at the time of first publication of this Guide (Linnhoff, 1982).

Experience in application has shown that it would be useful for process engineers to have an early appreciation of the implications on some of the network alternatives on the design of individual heat exchangers. For this reason, there is a section on "Heat Transfer Equipment" specially developed for the Guide. Simple technical descriptions of heat exchangers allow the non-specialist to decide on feasible types of heat exchangers for a given duty (plate, shell-and-tube, double-pipe, *etc.*) and to carry out cost comparisons in a convenient way between different types.

To complement both the description of the networking principles and the section on heat transfer equipment, outline case studies are presented in Section 4. These studies all relate to real-life projects, but have been simplified to present an outline description of the projects in question. The aim of these case studies is to give the user a "feel" for the overall approach when tackling a live project. Also, they help to demonstrate the nature of the interaction between network integration and process design in a broader sense.

The widespread application of a new technology usually presents organisational and cultural problems of overcoming inertia. In the present case, three such problems show up more clearly than others. First, the problem of organisational management to make good use of the new technology within the framework of existing design departments, project teams, time frames and corporate "campaigns". Second, the problem of introducing the new material in university teaching. Third, the problem of how to accomplish the "technology transfer" of the new techniques into design practice in general. The Guide addresses these problems in Section 6 under "Concluding Remarks".

What is Involved in Learning the Techniques?

The Guide is intended to be a self-teaching document. Studying Sections 2 and 3, solving the example problems and reading the outline case studies should take the user approximately five days of concentrated effort. Thereafter, he should be able to tackle his own problems generating better energy recovery networks.

However, a word of warning seems appropriate. Like most techniques based on concepts rather than rules, the techniques require a good understanding and some creative flexibility on behalf of the user. Without these assets the user will not be able to take full advantage of the generality and the flexibility offered by the techniques. An inkling of the type of *ad-hoc* arguments necessary when applying the techniques to specific projects can be obtained from the outline case studies in Section 4.

Naturally, there is a limit as to the amount of material that will fit into a document as compact as this Guide and readers wishing to extend their knowledge of the methods are advised to consult the original research papers as given in the list of references at the end of the Guide. Furthermore, short courses are an obvious aid to an in-depth understanding and appreciation of the tricks and subtleties involved in practical applications. At the time of first publication of this Guide, such courses are available at UMIST (Linnhoff, 1982).

1.1. Bringing the Guide up to Date

Through much of its fifteen year history pinch analysis has been somewhat controversial. Early on, as pinch technology, it introduced simple concepts in a field and an era otherwise known for complex mathematical methods. Reports from first industrial applications[37,42,74] claimed design improvements so large as to invite incredulity. Moreover, pinch analysis was commercialized early in its development when there was little know-how from practical application. This led to several commercial failures and many vested interests. To this day, opinions remain divided. Morgan (M. W. Kellogg) reports that pinch analysis significantly improves both the 'process design and the design process'[56]. Steinmeyer (Monsanto) is concerned that pinch analysis may miss out on '. . . major opportunity for improvement . . .'[75].

As is often the case when there is a difference of opinion, different parties have different data. Some of the concepts of pinch analysis are widely known, others are less well known. Experience with practical application varies. Some organizations have successfully integrated the approach into the design process, others rely on one or two technical experts. In addition, pinch analysis has undergone evolution making it difficult for the casual observer to come to an up-to-date assessment. It has evolved from the design of heat exchanger networks into a general methodology for the design of integrated systems ranging from processes to total sites.

This is a state-of-the-art overview. It gives a summary of key established concepts with special emphasis on those concepts which are less well known than others. This is followed by a description of eight recent developments. Some of these developments sharpen the focus of the established concepts, others widen the scope. All of them follow the established pattern of pinch analysis of setting targets prior to design. The eight developments are:

- Pressure drop optimization
- Multiple base case design
- Distillation column profiles
- Low temperature process design
- Batch process integration
- Water and waste water minimization
- Total site integration
- Emissions targeting

An overview concentrates on the essence not the detail. Case studies are described, and scope and impact of practical applications are assessed. For details of methodology, the reader is referred to the relevant references.

To complete the overview, there is a comprehensive summary of previous literature. Theory and applications are covered, and a historical perspective is given. The evolution of 'pinch' is discussed and it is argued that the term pinch 'analysis' is more appropriate today than pinch 'technology'. Pinch analysis is used for the scoping and screening of options during targeting prior to design. The choice of utilities, capital cost optimization, changes in the process/reactor/separation system, overall process emissions, etc. are all evaluated prior to design i.e. in analysis mode.

The Evolution of Pinch Technology and Pinch Analysis

In the late 1970s pinch technology emerged as a tool for the design of heat exchanger networks against the background of the then-current energy crisis. Its key contribution was to give the engineer simple concepts which were used interactively. The approach is that of the first edition of this Guide (the remaining pages of this book) and differed from most of the black box computer-based methods proposed at the time. Using the concepts of pinch technology the engineer could stay in control. He or she could use simple targets, take account of operability, plant-layout, safety, etc. and drive the design towards solutions which were not only thermally efficient (the input of pinch technology) but also industrially acceptable (the engineer's input).

In the mid 1980s Gundersen and Naess published a seminal review of heat exchanger network design methodology[23]. They gave prime attention to pinch technology emphasizing its contribution to the definition of heat exchange structures. This fairly reflected at the time the accepted view of pinch technology.

Since then, the methodology has become more broadly-based. Its principles are still based on those of heat and power thermodynamics and its key strategy is still to set targets prior to design (the analysis role of thermodynamics). However, the methodology has been extended to address systems including distillation, heat pumps, co-generating turbines, furnaces, etc. and to address non-energy objectives such as capital costs, operability, and emissions. A general approach has emerged for the integrated design of energy and process systems.

The feel of the methodology has changed. Its scope has widened and its emphasis has become targeting rather than design. In practice, it is now primarily used for 'process analysis' and is therefore increasingly referred to as pinch 'analysis' rather than 'technology'. Morgan explains that today's experienced practitioner will use pinch analysis to scope and screen alternative options early during the conceptual design of integrated systems[56]. In other words, the analysis is used as a general methodology for conceptual process design upstream from flowsheeting and simulation. It is no longer restricted to heat exchanger networks.

Key Established Principles—The Basics

Figures 1 and 2 relate to the three concepts probably best known in pinch analysis, the 'composite curves'[27], the 'grid diagram' of streams[34], and the 'pinch'[35]. All three concepts relate to a specific economic level of heat recovery which needs to be optimized (see opposite).

The composite curves are constructed from 'stream data' representing a process heat and material balance. The composite curves allow the designer to predict optimized hot and cold utility targets ahead of design, to understand driving forces for heat transfer, and to locate the heat recovery 'pinch'.

The grid diagram of streams allows the designer to develop heat recovery

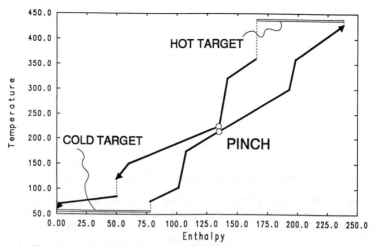

Figure 1—The best established concept of Pinch Analysis is that of Composite Curves. The Composite Curves set energy targets ahead of design and locate the Pinch point. They are also used to visualize heat transfer (sources, sinks, driving forces) in the overall context

PINCH

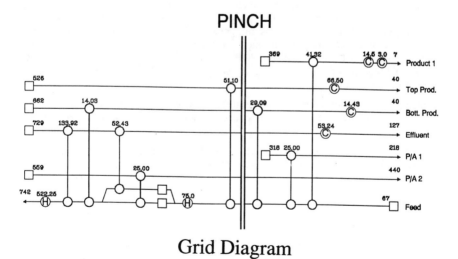

Grid Diagram

Figure 2—The Grid Diagram of streams is laid out prior to design. Hot streams run from left to right at the top, cold streams run countercurrent at the bottom. This grid does not change whatever design follows. The Pinch is easily located in the grid

networks using the 'pinch design method'[39], a mixture of methodology and engineering judgement.

The pinch principle states that no design can achieve the optimized utility targets if there is any 'cross-pinch' heat transfer from above the pinch to below the pinch:

$$\text{Actual} = XP + \text{Target}$$

(where XP = cross-pinch heat transfer). If $XP = 0$, then the actual heat requirement equals the target. The pinch principle allows the designer to keep the level of heat recovery optimized all through the evolution of even complex designs by simply making sure that cross-pinch heat transfer is minimised.

Figures 3 and 4 describe the probably next best known concepts, the 'grand composite curve' and 'appropriate placement'.

Grand composite curves (Section 2.2.5.1) are based on the same process stream data as composite curves. Grand composite curves highlight the process/utility interface. They help the designer select between different utility sources and sinks.

Appropriate placement applies to heat engines[79], heat pumps (Section 2.3.3.1), distillation columns[40], evaporators[70], furnaces[48], and to any other unit operation, be it process or utility, which could be represented in terms of heat sources and sinks. In Figures 3 and 4 the example is that of a heat engine. Appropriate placement for heat engines is above or below the pinch. Two turbine stages (1 and 2) above the process pinch consume 120 units of very high pressure (VHP) steam, deliver 100 units of steam (HP and MP) to the process and generate 20 units of shaftwork. Given that without a turbine the process would have required 100 units of heat in any case, the shaftwork generated above HP and MP steam is equal to the additional steam consumed. Stages 1 and 2 are *appropriately placed*. Essentially, shaftwork is generated from steam one-to-one.

Stage 3 below the process pinch generates 5 units of work, the heat equivalent of which would otherwise have passed to cooling water. This stage is *appropriately placed*, too.

Any turbine stage between input levels VHP/HP/MP and output levels VLP/condensate would be across the process pinch and would result in no more than the usual stand-alone turbine efficiency. Such a stage would be *inappropriately placed*.

The concept of appropriate placement holds true for any unit operation and any process. It gives the designer powerful guidelines for the design of heat engines, heat pumps, distillation columns, evaporators, etc. in the context of integrated processes.

There has been a lack of clarity in the literature about the role of the process pinch and that of utility pinches in appropriate placement. An example is reference 11. Please note that in Figure 4 work is generated above, below, or in between process and utility pinches, but not across *any* pinch, be it process or utility. It does not matter thereby whether the process provides the source or the sink. If work was generated across the process pinch overall hot and cold utility would increase. If work

A4

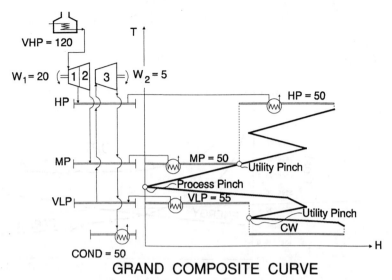

GRAND COMPOSITE CURVE

Figure 3—The Grand Composite Curve helps select utilities

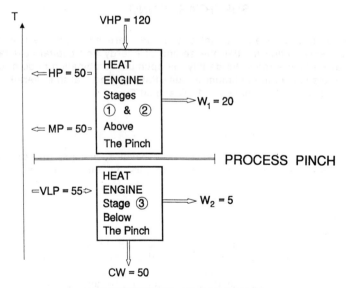

APPROPRIATE PLACEMENT

Figure 4—The concept of Appropriate Placement assesses the integration of individual process or utility units in the overall context. The example is based on Figure 3. The Appropriate Placement of heat engines is not across the Pinch

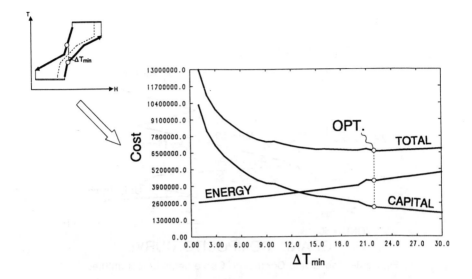

SUPERTARGETING

Figure 5—Both energy and capital cost targets are set from the Composite Curves. SuperTargeting implies the setting of energy and capital cost targets prior to design over a range to identify the optimum combination of both targets In the example shown the optimum is quite flat implying that the overall cost is similar for values of ΔT_{min} between 15°C and 30°C

was generated across a utility pinch, heat balance changes will cancel out but the utilities above and below the pinch in question could no longer be optimized.

Supertargeting
Figure 5 relates to the concept of 'supertargeting'[52]. It is possible to predict from the composite curves the overall target surface area for any stream data. It is further possible to predict the minimum capital cost necessary for the target area. Based on target energy cost and target capital cost, the minimum overall cost is predicted for any given ΔT_{min} (i.e. the minimum allowable temperature difference for heat exchange). This enables the designer to optimize ΔT_{min} *prior to design*. In other words pinch analysis does not just address energy cost. It is based on thermodynamics but it takes into account capital cost. It includes the overall optimization of energy *and* capital cost of integrated systems.

A6

Plus-minus principle

Figure 6 relates to 'process changes' as discussed in the literature (Section 5.1.3). This implies changes to the basic process heat and material balance but not necessarily to the chemistry of the process. The so called 'onion diagram' is used to represent the hierarchy of process design. We start with reactors turning feeds into products. Once the reactor design is settled, we understand the separation needs (for example, to separate unconverted feeds from products) and we can design the separators. Once the separator design is settled we understand mass flows, vapour/liquid equilibria, temperatures, pressures, and flowrates, and we can design the heat exchanger network. Prior to designing the heat exchanger network, we have defined the basic heat and material balance and we are able to construct composite curves representing the base case process.

The relevant concept for process changes is the 'plus-minus principle'[41]. This concept is essentially a generalization of appropriate placement. It uses the composite curves to identify process changes by going 'backwards' into the onion. The example shown in Figure 6 relates to the change of pressure in a distillation column. If column pressure is reduced then condenser temperature and reboil temperature will decrease. The condenser does not cross the pinch. Therefore, there is both a ⊖ and a ⊕ in the hot stream enthalpy balance below the pinch. The effects cancel out. There is no net change in utility resulting from the temperature change of the condenser. The reboiler by contrast was originally 'placed' above the pinch and crosses the pinch as a result of the pressure change. There is a ⊖ above the pinch and a ⊕ below the pinch in the cold stream enthalpy balance which results in both hot and cold utility targets being reduced. Essentially, the distillation column could now be run on integrated heat flow, or zero utility.

Process Changes

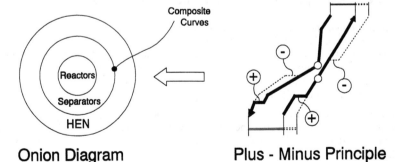

Onion Diagram

Plus - Minus Principle

Figure 6—Composite Curves are based on the reactor/separation scheme of the base case process. Composite Curves can therefore be used to modify the reactor/separation scheme. This task is guided by the Plus-Minus Principle

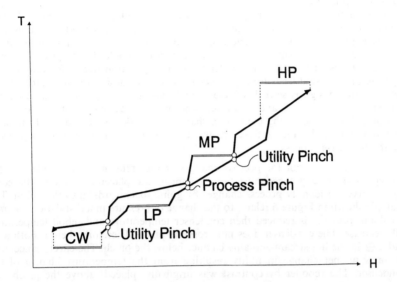

BALANCED COMPOSITE CURVES

Figure 7—The Balanced Composite Curves represent process and utility data such that the enthalpy balance is closed. The Balanced Composite Curves help locate Utility Pinches and allow the designer to visualize overall heat transfer process/process and process/utility

The plus-minus principle is a general formulation for desirable process changes (relating to distillation, reaction, etc.) which gives the designer guidelines as to what changes are beneficial for energy (as in Figure 6) and/or capital costs[45].

Balanced composite curves and balanced grid
Figure 7 shows a 'balanced composite curve'[45]. Balanced composite curves represent process and utility streams in one construction. Utilities are integrated with process streams so as to close the overall enthalpy balance. Balanced composite curves are desirable for a clear visualization of overall heat transfer (utility/process and process/process) and of utility pinches.

Figure 8 is a schematic of a 'balanced grid'[45]. The balanced grid consists of process stream data *and* utilities as per the balanced composite curves. Process changes and optimized values for ΔT_{min} as obtained through supertargeting are the basis of the balanced grid.

The stream and utility data used in Figures 7 and 8 are based on the example shown in Figure 3. It is interesting to note that the upper temperature utility pinch is caused by *and is located at* the utility. The lower temperature utility pinch

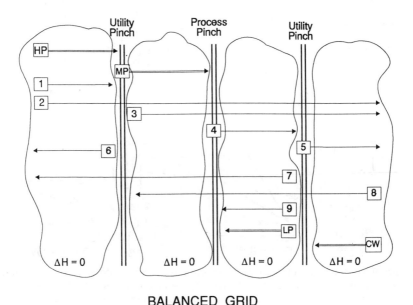

BALANCED GRID

Figure 8—The Balanced Grid combines process and utility data as per the Balanced Composites. Design carried out in the Balanced Grid will give equal consideration to process/process and utility/process units

is caused by utility *but is located at* the start of process stream No. 5. The concept of utility pinches is generally less well understood than other basic concepts. Full explanations are provided elsewhere[4 5].

Figures 9 and 10 demonstrate the usefulness of the balanced grid. Figure 9 shows two competing designs for a simple heat exchanger network. Both designs feature fired heaters with a coil duty of 300 units (the target, see Ref. 45). There is a difference, however. Process temperatures are different resulting in different flue gas temperatures and therefore different fired heater efficiencies. Even though both designs achieve the targeted heater duty, fuel consumption differs significantly (89% vs. 100%). In Figure 10, the balanced grid is laid out first, including a targeted flue gas profile (from 1500°C to 160°C). Normal use of network design procedures now identifies process/process exchangers as well as fired heater duties in one coherent approach. The resulting design achieves (a) the targeted heat duty and (b) the targeted minimum flue gas temperature. Fuel consumption is reduced to 81%.

The most common mistake made even by otherwise experienced users of pinch analysis is the consideration of utility choices *after* preliminary design. Design should be carried out based on balanced composites and in the balanced grid. This ensures

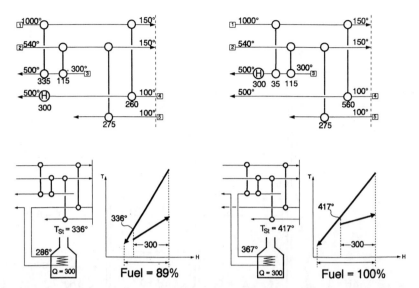

Figure 9—In this example, two HEN designs are shown both achieving 300 units of heat input (the energy target). However, the location of the heater is different between the designs. Therefore, the process temperature at heater input is different resulting in different furnace efficiency and different fuel consumption

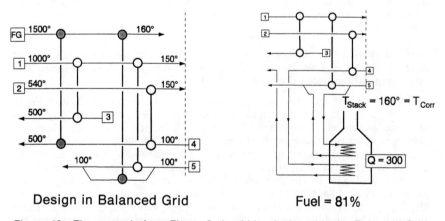

Design in Balanced Grid   Fuel = 81%

Figure 10—The example from Figure 9 should be designed in the Balanced Grid. This will ensure that (a) the energy target is reached and (b) the minimum flue gas temperature is reached also. Once the design is completed, the Balanced Grid representation is easily changed into a more conventional drawing which shows the intended use of the process furnace

it is carried out *after* proper targeting of utilities, consideration of process changes, and optimization. The balanced composite curves and the balanced grid are less well known than other basic concepts. They are, however, of key importance.

Remaining Problem Analysis (RPA)

Figure 11 describes the concept of 'Remaining Problem Analysis'[45]. This is the least well-known of the key established concepts. Let us assume the simple energy target for a given stream set is 100 units (Figure 11a). Let us further assume there is an operability constraint: a start-up heater is necessary on stream No. 3. To save equipment the designer wishes to place the steady state utility heater with the targeted heat load (i.e. 100 units) on stream No. 3, where the heater can also be used for start-up. In Figure 11b, the impact of such a decision is tested using Remaining Problem Analysis (RPA). The energy target for the 'remaining problem' is 20 units. This means that the heater on stream No. 3 is well located for start-up but not for effective thermal integration; if all 100 units of steady state heat load are placed on stream No. 3 an overall penalty will be incurred of 20 units. In Figure 11c a compromise is explored. It is possible to place 40 units of utility on stream No. 3 (for start-up *and* steady state load) without penalty.

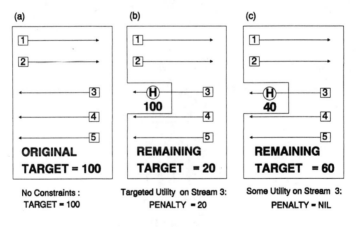

REMAINING PROBLEM ANALYSIS (RPA)

Figure 11—The target for the unconstrained problem is 100 units. If this target heat duty were to be placed on stream No. 3 (for operability reasons), then Remaining Problem Analysis (RPA) would reveal that there is now a target of 20 units for the remaining problem. Thus overall consumption would be 120 units; there would be a penalty of 20 units. If only 40 units of utility are used on stream No. 3 then RPA reveals that there is a target of 60 units for the remaining problem. In other words the designer is free to use 40 units of utility on stream No. 3; there will be no penalty

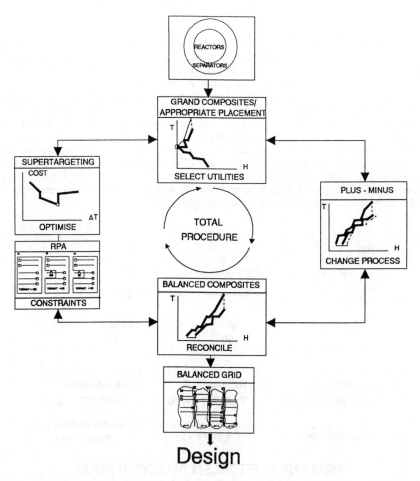

Design

Figure 12—Pinch Analysis is the scoping and screening of design options in targeting mode prior to design. Utility selection, capital cost optimization, the impact of constraints and of changes to the process/reactor/separation system, are all evaluated and put into context. The Balanced Composite Curves and Balanced Grid represent an optimized set of targets. Design only proceeds after this initialization

RPA can be carried out with respect to both energy *and* capital cost targets. It is a powerful evaluation tool allowing the designer to explore the impact of key design decisions during targeting. Design features such as 'a start-up heater on stream No. 3, no larger than 40 units' are easily agreed at an early stage and form the basis of subsequent design work.

The overall approach
Figure 12 summarizes traditional pinch analysis. We start with the base case heat and material balance representing reactors and separators. The grand composite curves are constructed and an initial assessment is made of utilities. Supertargets are set and/or process changes are examined. Utility options are tested again. Remaining Problem Analysis is used to evaluate the impact of constraints. Capital costs are traded off against energy costs. Based on these targeting exercises, assumptions are varied regarding key process design options (reactor and separator design), key utility scenarios (e.g. fluegas vs. steam), key process constraints (e.g. start-up heaters), and major capital cost trade-offs. At the end of this iterative targeting procedure the designer will have decided on a basic design strategy. He or she will have selected the correct broad scenario and initialized the design near-optimally. The design of the heat exchanger network now follows (in the Balanced Grid), as does the design of the main process items and utility units.

Figure 12 makes it once more appropriate to reflect on the terms pinch 'technology' versus 'analysis'. When 'pinch' first emerged, the grid diagram and the design rules for heat exchanger networks were the main part of it. Network design was its key contribution. Targeting was secondary. By contrast, *the key contribution today is the broad scoping and screening of process and utility scenarios at the targeting stage.* It would therefore appear more appropriate today to refer to 'pinch' as an analysis tool rather than a technology. This should explain why workers in the field and users alike increasingly use the term pinch analysis in preference to pinch technology. The present paper uses pinch analysis consistently, except where this is not appropriate in the historical context.

Finally, the procedure described in Figure 12 clearly contradicts the widespread view that pinch analysis is simply a tool for the design of heat exchanger networks. Rather, pinch analysis has evolved into a general methodology for overall process design.

Further Literature

The present paper can only give a description of a small number of principles and concepts of pinch analysis. In all, there are some sixty principles and concepts. Table 1 gives a near-complete summary. Not all of these concepts are covered in the literature quoted so far. Most of the remaining concepts are covered either in Linnhoff March's 'Foundation Training Course'[45] or in UMIST's 'MSc Course in Process Integration'[80]. In addition, Table 1 includes further references.

For the reader who wishes to understand the essence rather than the detail, there are previous overview papers from ICI[5,36,37,78], ETSU[6], BASF[31,46], Union

A13

Table 1—Principles, tools and design rules of pinch analysis. The numbers relate to the references which either introduce or explain the concept in question

Principles and tools	Rules

Basic hen design

Composite curves[278] (Section 2.2.1)   Pinch design method[39]
Problem table[34]
Threshold plot[35]   Stream splitting rules[45]
Bath area target[52]
Delta-T contributions[52]   Mixing rules[80]
Euler & no of units target[34]
Grid diagram[34]   HEN evolution[45]
Pinch principle[35]
CP matrix[45]   Topology traps[52]
Criss cross principle[52]
Driving force plot(s)[80]   Retrofit design[80]
Remaining problem analysis[45]
Loops & paths[34]
Supertargeting[52]
Delta p targeting and supertargeting[62]
Retrofit targeting (constant α)[77]
Area matrix[67]

Utility targeting and design

Grand composite curve   Furnace integration[48]
(Section 2.2.5.1)
Utility composite & utility pinches[45]   Heat engine placement[79]
Balanced composites & balanced   Heat pump placement[79]
grid[45]
Total site profiles $(T - H)$[14]   Multiple utility optimization[61]
Cooling water targets[80]   Low T process design[15]
Exergy & pinch: the $\eta_c - H$ plot[50]
Exergy grand composites[15]
Low T shaftwork target[53]
Total site profiles $(\eta_c - H)$[14]
Power cycle targeting[16]

Advanced hen design

Rigorous area targets[66]   Constrained HEN design[59]
Constrained HEN targeting[8]
Area matrix retrofit targets[67]   Condensing steam cycle power block
Area integrity matrix[3]   design[51]
Area cost targets for different
materials of construction[24]
No of shells targets[2]
Downstream paths[44]
Sensitivity tables[32]

Table 1—Continued

Multiple base case targets[28]
Repiping & rerouting targets[28]
Non-convexity[7]
Resiliency index[67]
Time slice targets[58]
Cascade analysis[30]
Batch utility curves[22]

Process design

The onion model (Section 6.3)	Data extraction rules[45]
Keep hot streams hot and cold streams cold (Section 2.2.7.2)	
Plus/minus principle[41]	Appropriate placement of distillation columns[40]
Column grand composite curves[13]	
Column composite curves[13]	Appropriate placement of evaporators[70]
Waste water targets[81]	
	Waste water system design rules[81]

Carbide[42,43], UMIST[47,68], Linnhoff March[49], M. W. Kellogg[56] and Mitsubishi Kasei[57]. These papers summarized developments from time to time and reported on impact and implications. Some of them offer the most convenient access to pinch analysis available to date. One paper not only gives an industrially oriented overview but very interestingly discusses the way in which pinch analysis should be integrated into the design process[56]. This is an important subject. An industrial organization wishing to take on board pinch analysis must look beyond the training of experts. It must look at information flow, at schedules and the organization of project work, the spreading of general awareness, and other aspects of the organization.

The remainder of the overview outlines eight selected areas of recent progress. The discussion will centre on application, scope, and impact. For details of methodology, the appropriate references are given. For five of the developments these references originate from the Department of Process Integration at UMIST. For the remaining three developments, there are references from other sources as well. Clearly, UMIST has played a dominant role in developing the concepts of pinch analysis over the years. It is therefore promising that some of the more recent developments originate from other centres also.

Pressure Drop Optimization

Figure 13 is a sketch representing a crude oil preheat train design and highlighting the furnace as well as the feed pump. Furnaces and pumps are often limiting in the revamp or debottleneck of crude preheat trains. The challenge is to make the unit more fuel efficient and debottleneck the furnace by adding and reconfiguring surface area *while at the same time avoiding the installation of new pumps*. In other

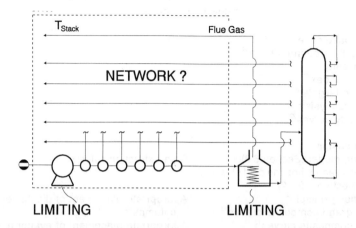

Figure 13—Heat recovery systems consume both power and fuel. Often either the pumps or the furnaces are limiting

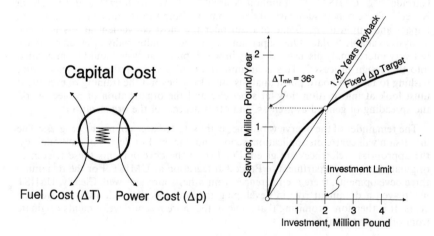

Three-way Trade-off   Three-way Targeting

Figure 14—Surface area in heat exchangers is optimized (1) against temperature difference, i.e. fuel and (2) against pressure drop, i.e. power. There is a three-way trade-off. Pressure drop targeting considers this three-way trade-off either by allowing pressure drop to be optimized (new design) or by accepting given pressure drop limits (retrofit)

words, revamp or debottlenecking objectives need to fit in with pressure drop limits set by existing pumps. This is a complicated design problem.

Traditional pinch analysis ignored pressure drop. Heat exchangers were designed in the context of temperature and heat load considerations. Pressure drop was considered as an afterthought. It became apparent during the first practical applications of pinch analysis that pressure drop could not be treated in this fashion. An optimized network would settle heat exchangers at given sizes. Subsequently, optimization of heat exchanger surface area against pressure drop might double certain exchangers in size, while others became much smaller. This clearly invalidated any optimization that had taken place during the initial design. Worse, in retrofits, the initial design would often exceed the available pressure drop limits rendering the design impractical. The conclusion was soon reached that the optimization of heat exchanger surface area vs. thermal energy is inextricably linked with the optimization of heat exchanger surface area vs. pressure drop. There is a three-way trade-off.

This problem led, after several years of work, to the concept of 'pressure drop targeting' and of 'three dimensional supertargeting', see Figure 14[62]. Three dimensional supertargeting takes on board pressure-drop related costs or pressure-drop limits for hot and cold streams. Targeting is then carried out consistent with the cost of (or limits on) pressure drop, the cost of fuel and the cost of heat exchanger surface area. Prior to design, streams are set at optimized pressure drop, or at the available pressure drop limit, and utilities are set at the optimized level of thermal recovery. Heat exchangers are placed in this context. If an attempt is made to optimize a design so initialized for fuel, capital cost, or pressure drop, little benefit is found. Pressure drop targeting and three dimensional supertargeting have made an important contribution in making pinch analysis more practical, particularly in the context of oil refinery applications.

Multiple Base Case Design

Another subject proving important during the first practical applications of pinch analysis was that of flexible designs, or design for multiple base cases. Few designs, if any, are operated as per base case data. Processes need to operate efficiently, reliably and safely for different capacities, different product specifications, different feedstocks, fresh or spent catalyst, varying ambient temperatures, clean and fouled equipment, etc. The first industrial users of pinch analysis were sceptical as to the flexibility of integrated designs. The common thinking was that integration would lead to operability problems.

Initially, this hurdle was overcome only by hard work. Integrated structures had to be evolved for the base case and operability had to be checked in the traditional way, i.e. through simulation, modifications to the design, more simulation, etc. Experience showed quickly that while there was a relationship between integration and operability, there was not necessarily a conflict. In some cases, a given integration feature could prove beneficial for operability. In other cases, the same feature could be detrimental.

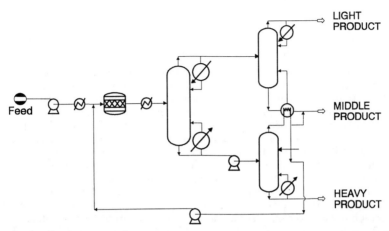

Figure 15—In this example, integration of the column condenser/reboiler may have beneficial or detrimental implications on operability. If catalyst deactivates and reactor product composition changes, the integration may prove detrimental. If overall throughput increases, and the utilities become limiting, the integration may prove beneficial

Consider Figure 15. A product separation/recycling system consists of three distillation columns. The condenser of the heavy products column is integrated with the reboiler of the light products column. Now consider two potential changes to operating conditions: (1) change of catalyst performance and therefore reactor product composition, and (2) change of feed flowrate. If the catalyst deteriorates and the reactor product composition changes the proposed integration may prove detrimental to operability. The load on one column may increase while that on the other column may decrease and the condenser/reboiler integration is likely to bottleneck capacity. If, by contrast, the overall feed flowrate changes the proposed integration may prove beneficial. A given variation in overall process capacity would result in a *smaller* variation in utility loads as a result of integration and in a case where the utility system is limiting this may prove beneficial to debottlenecking the overall process.

The overall experience today is that integrated systems can be more operable than their less integrated counterparts provided operability is taken into account early during design. The approach taken in pinch analysis is to include operability objectives in targeting and during the development of the integrated structure.

An example of this approach has already been discussed in the context of Remaining Problem Analysis (RPA). In Figure 11, RPA was used to settle the issue of a start-up heater *prior to design*.

Another example is referred to in Figure 16[77]. Different operating cases are examined in terms of their target curves (for overall surface area vs. energy) and in terms of their predicted energy consumption as per simulation. It is found that the

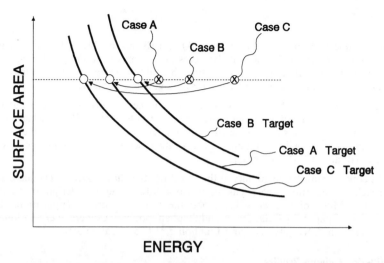

ENERGY

Figure 16—In multiple base case design targets are set for each operating case. Performance of a design is then assessed for each case relative to its target. In the example shown the design is reasonably well suited for Cases (A) and (B). It is not suited for Case (C)

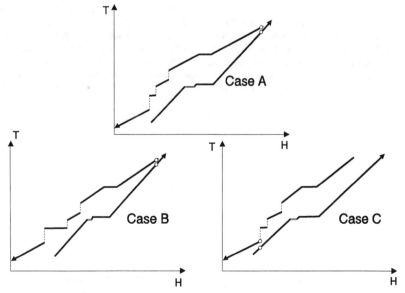

Figure 17—Composite Curves make it obvious why the example from Figure 16 shows good compatibility between Cases (A) and (B), but poor compatibility with Case (C)

design in question suits operating cases (A) and (B) reasonably well but does not suit case (C). Additional costs will be necessary to make the design flexible with respect to case (C). Alternatively, the mismatch could be accepted. Case (C) may not represent a frequent case and poor efficiency might be acceptable.

Figure 17 explains why certain designs may suit certain operating cases but not others. Simple study of the composite curves reveals that a design which suits case (A) is likely to also suit case (B) with few additional features but that more significant changes will be required to operate cases (A) and (C) in one design or cases (B) and (C) in one design.

The consideration of operability at the targeting stage has proved a very successful approach. In addition, there has been work addressing the design of flexible structures. Much of this work has stood the test of industrial application. Key references are papers by Calandranis and Stephanopoulos[7], Colberg et al.[10], Floudas and Grossman[20], Kotjabasakis and Linnhoff[32,33,44] and Saboo et al.[65,66].

Distillation Column Profiles

Moving on to subjects relating to the design of the basic process rather than the heat exchanger network, a breakthrough was achieved recently with regard to distillation. Consider Figure 18. On the left, the traditional pinch analysis view of distillation is shown. A so-called 'box' in temperature-enthalpy co-ordinates describes the reboiler as a heat sink and the condenser as a heat source at lower temperature. This is contrasted on the right with a graph between reboil and condensing temperatures not unlike a grand composite curve called a 'column profile'. The pinch point of the column profile is located at the column feed. Column profiles are used like grand composite curves. They indicate at what temperature heat needs to be supplied and rejected up and down the column. Not all heat needs to be supplied at reboil temperature. Some can be supplied at lower temperature. Likewise, not all

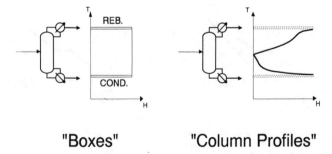

"Boxes"   "Column Profiles"

Figure 18—Traditionally, Pinch Analysis described distillation through the thermo-dynamic 'footprint' of the condenser and the reboiler in the so-called box representation. More recently, Column Profiles are used instead

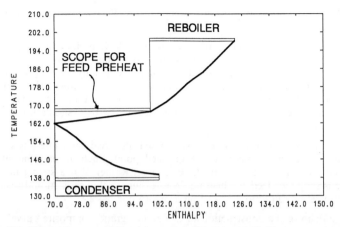

Figure 19—The column profile of a petrochemical distillation suggests that additional feed pre-heat would lead to a significant reduction in reboil duty

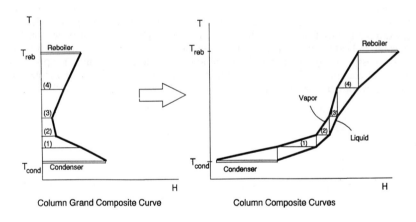

CAPITAL COST

Figure 20—Column Grand Composite Curves and Column Composite Curves have a similar relationship as established Grand Composite Curves and Composite Curves. Column Grand Composite Curves help assess the use of external heat sources and sinks. Column Composite Curves help interpret column internal processes, driving forces, and capital costs

heat needs to be removed at condensing temperature. Partial heat removal at higher temperatures may be appropriate.

Graphs such as column profiles are not new. They have been described previously in the literature[29,21]. However, in previous work such graphs were described theoretically. They could not realistically be computed for anything but ideal binary mixtures. Their computation for real columns was near impossible. Significant progress was made recently insofar as graphs of this type can now be generated with reasonable accuracy from a single converged tray-by-tray simulation even for multi-component non-ideal mixtures[13].

Figure 19 shows typical column profiles based on a real case. There is scope for significant heat supply directly above the feed point. Such an extreme situation usually does not call for an intermediate reboiler. Rather, it is a strong indication that there is scope for feed pre-heating. Pre-heating, in this case, should result in a reduction in reboil duty more or less on a one-to-one basis.

Figure 20 shows the construction of a 'column grand composite curve' and of 'column composite curves' in a tray by tray fashion. The column composite curves describe vapour and liquid travelling up and down the column and depict available driving forces. They are a strong clue to column capital costs. As with traditional pinch analysis, column composite curves allow the designer to include the consideration of economics (reflux ratio) hand in hand with technical feasibility.

Finally, Figure 21 contrasts the traditional approach of describing distillation columns as boxes with Column profiles in the context of an integrated process design. The example is based on three columns, A, B and C. The overlap of boxes (on the left) would indicate that for integration, the pressure in column A should be increased and the pressure in column C should be reduced. Consideration of the same problem in terms of column profiles (on the right), makes it apparent that a side reboiler in column B would enable integration to take place between columns A and B and a

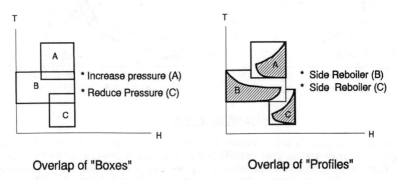

Overlap of "Boxes"   Overlap of "Profiles"

Figure 21—An integration problem is assessed using the traditional box representation and the more recent concept of Column Profiles. The conclusions are different. Column Profiles offer a better assessment

side reboiler in column C would be appropriate for integration between columns B and C. Clearly the conclusions are different. The latter conclusions are the valid ones.

Column profiles have proved a powerful tool over the last couple of years in their first industrial applications, both in the optimization of individual columns and in the re-configuration of integrated distillation schemes. However, there is work left to be done to further develop these concepts. This work should prove rewarding. Distillation is without doubt the single most important aspect of process design for energy efficiency. Designs for the future must either avoid or improve distillation. Column profiles help the designer with both objectives.

Low Temperature Process Design

Figure 22 is based on the example of ethylene process design. The process extends significantly below ambient temperature with several major distillations carried out either across or below ambient. There are many heat sources and heat sinks below ambient in the process with numerous integration opportunities. A complex refrigeration system supplies and removes heat to and from the process below ambient. The refrigeration system will usually consist of two cycles (ethylene and propylene), and will operate at several levels. There are only two utilities: cooling water and shaftwork.

The overall design of systems such as this is complex. There are the issues of both process design and refrigeration system design. Finally, there is integration between the process and the refrigeration system. We have a simultaneous design problem of

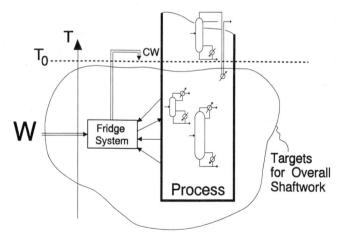

Figure 22—Low temperature separation processes such as encountered in ethylene production bring together the design of low temperature process units, complex heat recovery networks, and multi-level refrigeration systems. There are only two external utilities: shaftwork and cooling water

low temperature distillation, of other process operations, of process heat recovery, and of the refrigeration system.

Traditional pinch analysis could help in this context only to a limited extent. The usual approach was to use pinch analysis to set process heat load targets, to identify refrigeration levels and loads from these targets, and to design the fridge system subsequently. However, any change in process design then had to be translated into new refrigeration levels and loads and the design of the refrigeration system needed to be updated accordingly. The ultimate effect of any change in the process on capital costs and on power consumption was therefore evaluated by means of a somewhat cumbersome procedure.

Recent work has led to shaftwork targets for overall low temperature systems[53]. Consider Figure 23. A process grand composite curve is shown below ambient with the Carnot Factor as vertical axis instead of temperature. Due to this substitution, the area in the construction represents exergy[50]. Specifically, the cross-hatched area between the process grand composite and the refrigeration levels represents exergy loss. Consequently, a change in refrigeration system design as shown in Figure 23 (the example relates to the introduction of an additional level) is easily assessed in terms of the consequent reduction of the exergy loss and therefore of the exergy supplied by the refrigeration system. The reduction in overall shaftwork is:

$$(W_A - W_B) = [1/\eta_{ex}](\alpha)$$

where η_{ex} (the exergetic efficiency of the refrigeration system) is approximately constant.

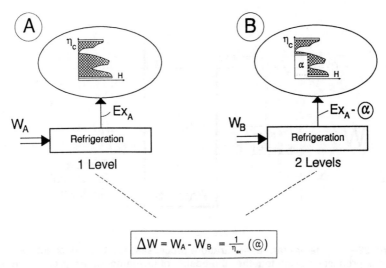

Figure 23—The change from a single level refrigeration system design to a two-level system saves shaftwork proportional to area α

A24

DESIGN Ⓐ

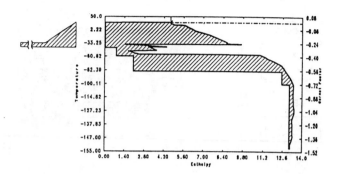

DESIGN Ⓑ

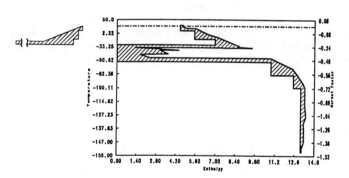

$$\Delta W = \frac{1}{\eta_{\theta x}} (\blacksquare - \blacksquare) = \frac{1}{\eta_{\theta x}} (\ 2.26 \ MW \ )$$

$$= 3.83 \ MW$$

Figure 24—In this example based on ethylene process design shaftwork targets are used to predict the benefits to be obtained from changes in the heat recovery and in the refrigeration system[53]

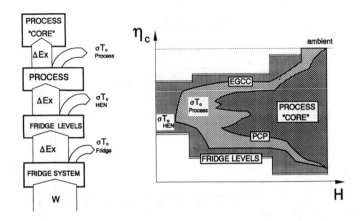

Figure 25—The so-called Exergy Grand Composite Curve applies Pinch Analysis to Carnot Factor/enthalpy axes. Area represents exergy or exergy loss. The innermost profile surrounds area representing the process 'core', i.e. the ideal process exergy requirement. The next outer area represents process losses. The next outer profile represents the exergy requirement of the non-ideal process. The next outer area represents heat recovery losses, and the next outer—stepped—profile represents fridge levels. This type of construction lets the designer deal with process changes, the heat recovery network, and the design of the fridge system simultaneously both in terms of exergetic assessment and Pinch Analysis[15]

In other words, it is now possible to completely by-pass the design of the refrigeration system in targeting. Use of the approach is demonstrated in Figure 24[53]. A base case design (case A) is compared with an alternative design (case B). Cases A and B differ in terms of heat exchanger network and refrigeration system design. The shaftwork target predicts an improvement in overall power consumption of $\Delta W = 3.83$ MW. Detailed design and simulation of cases A and B identify an improvement of $\Delta W = 3.76$ MW. This implies a discrepancy of 1.9% between simulation and the targeting approach.

The approach extends to process changes. Consider Figure 25. The innermost profile, the so-called 'Process Core Profile' (PCP), surrounds the area representing the process exergy change due to reversible key process operations. The area situated immediately outside the PCP represents the exergy loss in process operations, $\sigma T_{oProcess}$. The next outer profile, the so-called 'Exergy Grand Composite Curve' (EGCC), represents the (non-ideal) process. The next area out represents losses in the heat exchanger network, σT_{oHEN}, the next outer profile represents the fridge levels, etc. The engineer uses constructions such as Figure 25 to consider process changes, heat exchanger network design, and refrigeration system design in one coherent approach[15].

The combined benefit of distillation column profiles and of low temperature shaftwork targets in the design of low temperature distillation based processes (such as ethylene) has been significant. Results achieved with these techniques offer significant improvements over results achieved using the previously established principles of pinch analysis[56].

Batch Process Integration

There is designated and there is multi-purpose batch production, and there are cyclic and there are random batch processes. Designated and cyclic production is often found in the food and drink industry. Multi-purpose and random production is found, e.g., in the manufacture of glue and resins, of pharmaceuticals, and of other low volume specialist products. The concept of integration would appear rather more obvious in the context of designated and cyclic production than in the context of multi-purpose and random production. Experience nevertheless suggests that the integration of batch processes can be highly worthwhile even in the least likely environment. In a summary report issued by the Energy Technology Support Unit (ETSU) in Britain the results were published of twenty-six practical applications of pinch analysis across various sectors of industry[6]. Eighteen of these projects related to batch or partial batch production ranging from designated/cyclic to multi-purpose/random. Some of the most worthwhile results were achieved in multi-purpose/random batch environments.

The results in question usually come from better energy management but are often not primarily aimed at energy cost reduction. They are aimed at debottlenecking, improved yields, more consistent product quality, less re-work, reduced product cycle-times, etc.

Batch processes are usually controlled by a combination of the following four parameters:

- material flow
  (e.g. waiting for next charge)
- heat flow
  (e.g. waiting to reach temperature)
- equipment capacity
  (e.g. waiting for next empty vessel)
- labour
  (e.g. waiting for next shift)

These parameters are typically interlinked, and heat flow plays a dominant role for technical feasibility. Material flow may be controlled by the rate of heating the charge or cooling the product. Equipment capacity may be controlled by reactor residence time, which in turn is controlled by heat transfer. The quality of separation

A27

may be limited by cooling temperature limits, etc. Careful energy management usually results in cost savings for

- capacity (debottlenecking),
- yield and product quality, and
- energy

in that order[6].

So, the obvious question is: how does the engineer take account of the time dimension? There are various concepts and procedures. There are 'time average models', 'time slice models' and 'time event charts'[58]. There is 'cascade analysis'[30], and there are 'batch utility curves'[22]. An overall summary of today's state-of-the-art procedures was compiled recently[4].

A simple example of successful batch process integration is given in Figures 26 and 27[22]. Figure 26 shows the flowsheet of a speciality process operated in cycles but in the context of overall random production and in multi-purpose reactor vessels. Figure 26 also shows the time event chart for the process in question. The cycle time is 5.6 hours and reactor No. 2 is limiting. Figure 27 shows the flowsheet and the time event chart for the project proposal. Reactor No. 1 has been debottlenecked through external feed heating while charging occurs. Reactor No. 2 has been debottlenecked through external feed heating *and* product cooling, both steps being integrated for heat recovery, but de-coupled time wise through the introduction of a simple storage tank. Cycle time is reduced to 3.25 hours. Capacity is increased from 100% to 172%.

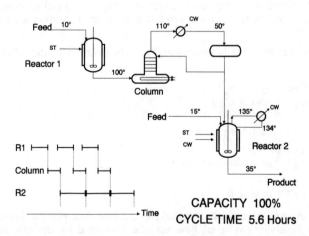

Figure 26—Time event chart and process sketch of a simple batch operation. Reactor No.2 is limiting[22]

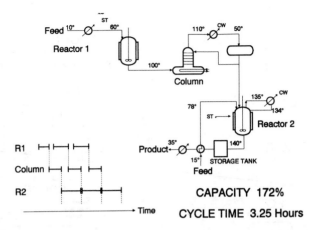

Figure 27—Project proposal for the system from Figure 26. Both reactors have been debottlenecked through external heat transfer (charge and discharge). Pre-heating and product cooling on reactor No. 2 are integrated but decoupled time-wise through a storage tank. Overall system capacity is increased by 72%[22]

As stated above, process integration may not be an obvious concept in the context of batch processing. There is, however, a significant body of commercial applications experience and, surprisingly, cost savings as a percentage of total operating costs are often greater than those found in continuous processes. In the author's opinion, the key reason is the fact that, in many batch process environments, there has simply never before been a *methodical* search for process integration opportunities.

Water and Waste Water Minimization

Water and waste water minimization are generic problems of integrated process design. Most (waste) water systems consist of a number of steps which relate to different parts of the process and which are linked by water. Water may be supplied as fresh water or may originate from the process. An example is given in Figure 28 and a simplified schematic involving five process steps is shown in Figure 29. In the schematic, each step requires water, for filter cake washing, steam stripping, extraction, etc. Fresh water is fed to step 1 and spent water is piped from step 1 to waste. Likewise, fresh water is supplied to step 3. However, spent water from step 3 is used in step 5 before it is passed to waste. In other words, water is 're-used'. Re-use results in a reduction of overall water flowrate. Finally, steps 4 and 2 are linked

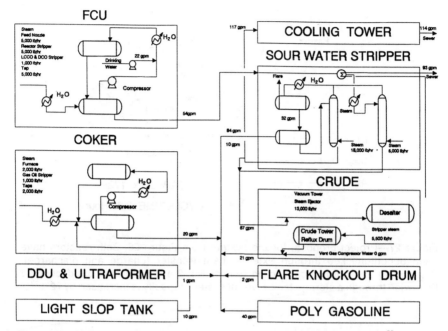

Figure 28—Example flowsheet of a refinery-based waste water system[55]

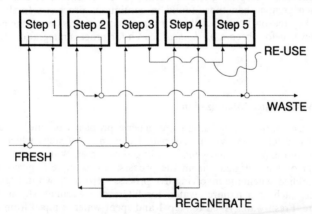

Figure 29—Schematic showing the logistics of waste water system design. Water is either used, re-used, or regenerated and re-used before it goes to waste. Re-use and regeneration lead to a reduction in overall water flowrate

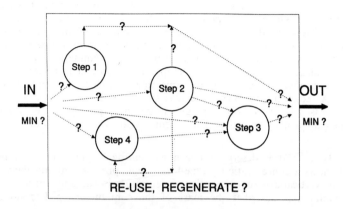

Figure 30—The correct structure of waste water systems incorporating re-use and regeneration sets a complex integration task

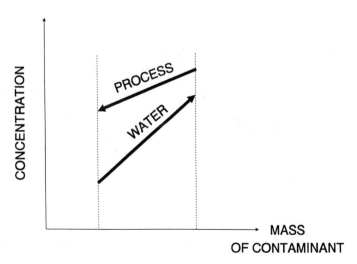

Figure 31—Each process step requiring water is represented in the co-ordinates of concentration and mass (of the picked up contaminant)

through regeneration. Regeneration is re-use with partial treatment prior to re-use. This also results in a reduction of the overall water flowrate.

Figure 30 describes the problem of integrated (waste) water system design. There may be any number of process steps, and water can be re-used and/or regenerated in all manner of possible configurations. As a result of re-use and regeneration, the overall water flowrate through the system is minimized. However, in practical applications there will be constraints. For example, water from one step may not be used in another step because of the danger of process contamination, etc. Even given these constraints, the overall task of identifying the correct steps for re-use and regeneration is often complex. At the same time, reduction of water flowrate is important. A reduction in flowrate decreases both fresh water costs and effluent treatment costs. Energy costs (pumping), capital costs (piping) and environmental load factors all usually improve. The incentives for water minimization can be strong.

Figures 31, 32, and 33 describe a relevant recent development in pinch analysis. Figure 31 shows a concentration profile for a single process step and the water to be used. The vertical axis represents the concentration of contaminant in the water. The horizontal axis describes the mass of the contaminant picked up by the water.

Figure 32 shows the water concentration profiles obtained for three individual process steps combined in a so-called 'composite concentration curve'. The rules of construction[18] are analogous to those for composite curves in temperature

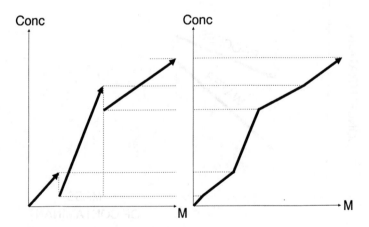

Figure 32—The construction of Composite Concentration Curves is analogous to that of the established Composite Curves in temperature enthalpy co-ordinates

CONC

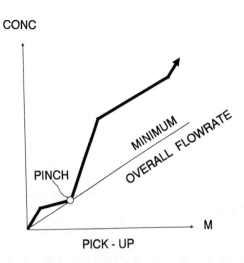

PICK - UP

Figure 33—The best possible tangent on the Composite Concentration Curve represents the minimum feasible water flowrate for the overall system. This sets the (waste) water target

enthalpy terms in traditional pinch analysis. Figure 33 shows a straight line drawn through the extreme point of the concentration profile (the 'pinch'). Clearly, a straight line of a given slope represents a specific overall water flowrate. For example, a straight line of steeper slope would represent a greater concentration change for a given amount of picked up contaminant, i.e. a lower water flowrate. The slope of the line obtained in Figure 33 determines the minimum feasible overall water flowrate for any system[81]. We have a minimum water target.

We have, again, a target which is set prior to design. We are able to predict the minimum feasible water flowrate *of any system*. However, reference 18 goes further. The pinch point in Figure 33 is significant. There are rules for the design of the integrated water system. These rules are analogous to those obtained for heat exchanger network design. There must be no concentration mixing across the pinch, etc. A methodology follows for the design of integrated process water systems which achieve the minimum water target.

Figure 34 demonstrates the design evolution for our example. Schematic (a) shows the simple parallel arrangement for steps 1, 2, and 3. The water flowrate would be 100%. Schematic (b) follows directly from the composite concentration curve and pinch design rules, and schematic (c) and (d) are simple evolutions. Water flowrate is reduced by 39%. The key feature in schematic (d) is the by-pass around step 1.

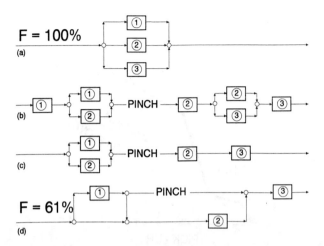

Figure 34—Design evolution for the example from Figures 32 and 33. Schematic (a) represents the simple 'parallel' solution. Schematic (b) follows directly from the Composite Concentration Curves. Schematics (c) and (d) represent evolutions. Note the design has been 'relaxed' in the areas away from the Concentration Pitch. The final evolution offers a 39% reduction of overall water flowrate at the expense of changes in piping

These concepts have recently been extended to apply for multi-component systems, (i.e. systems with more than one contaminant) and to the optimal placement of regeneration units[81]. Current development aims to extend the concepts further to the optimization of trade-offs between energy/capital/water costs/emissions and to the design of process waste water systems in the context of the overall factory. Distributed effluent water treatment can be beneficial if there are different process waste water streams from different processes. The emerging technology addresses the task of optimized water use within processes simultaneous with either distributed or centralized treatment plant design in the factory as a whole.

Total Site Integration

Total Site Integration takes us back to the roots of pinch analysis, i.e. to the optimization of heat and power. Different to traditional pinch analysis, however, total site integration takes us beyond the boundaries of the individual process. Consider Figure 35. There is the central combustion of fuel, there may be central cogeneration of power, and there are usually central steam levels. The production processes consume steam from the central mains. Some processes generate steam and others may burn fuel locally. total site integration does not address direct interactions

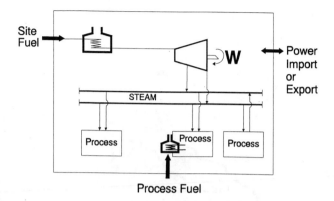

Site Fuel

STEAM

Power Import or Export

Process   Process   Process

Process Fuel

Figure 35—Schematic of a Total Site. Each process needs to be designed in the context of the factory's infrastructure. In turn, the infrastructure needs to optimally serve the processes. There are usually only three major components to overall energy costs: centrally fired fuel, fuel fired in the processes, and overall site power

between processes. Rather, it addresses the task of designing each process optimally in the context of the overall factory infrastructure.

A recent development is the concept of 'total site profiles'[14]. Figures 36, 37, and 38 describe the concept by way of a case study[54]. Figure 36 shows the utility system sketch of an existing factory with six processes. There is a high pressure (HP) mains which receives steam from the central cogeneration set. There is a medium pressure (MP) mains which recovers heat from certain processes and supplies heat to other processes, etc. Next to the sketch, the same site is described through total site profiles. The profiles are made up from the grand composite curves of the processes on site[14]. There is a sink profile and a source profile. The vertical axis is the Carnot Factor, the horizontal axis is enthalpy. Based on similar thermodynamic arguments as used in Figures 24 and 25, the use of the Carnot Factor in the vertical axis results in the fact that the shaded area in Figure 36 is proportional to shaftwork cogenerated on site.

There is a one-to-one relationship between the utility system sketch in Figure 36 and the construction based on total site profiles. There are 282 units of fuel consumed, 23 units of power cogenerated, 213 units of very high pressure (VHP) steam generated, 190 units of high pressure (HP) steam, 30 units of medium pressure (MP) steam, and 240 units of cooling water. The total site profiles are a thermodynamic representation of the energy flows on site, given the existing processes and infrastructure.

Figure 37 describes a proposed expansion plan. A new process is to be built on site. The utility system sketch shows that the new process will consume 70 units of MP

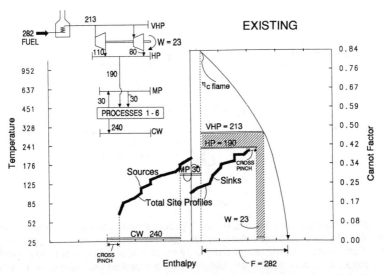

Figure 36—The existing site. There are six processes and a conventional energy system incorporating co-generation and two steam distribution levels (HP and MP)

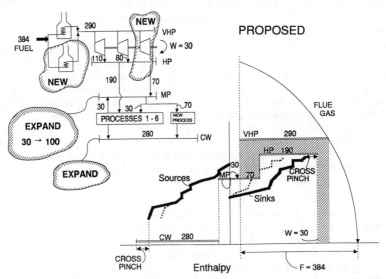

Figure 37—The proposed expansion. A new process is to be built on site. Substantial investment is required in the infrastructure. The project provides for additional cooling capacity, expansion of the MP steam mains, a new turbine (VHP/MP), and an additional furnace

steam. That steam is to be supplied via a (new) turbine which links into the existing VHP and MP systems. In order to generate the additional VHP steam required, a new boiler is planned. Also, an expansion of the cooling system is necessary (from 240 to 280 units).

Also in Figure 37, the total site profiles describe the same expansion plan. Again, there is a one-to-one relationship between the utility system sketch and the total site profiles. There are 384 units of fuel, 30 units of cogeneration, 290 units of VHP steam, 190 units of HP steam, 100 units of MP steam, and 280 units of cooling water.

However, the total site profiles reveal additional information. Clearly the pressure level of MP steam is not appropriate. As a result of the new process being added to the existing site, the process profiles have changed significantly. Importantly, the distance has increased between steam levels and most of the *existing* process sinks. It is apparent that MP steam does not exploit the driving forces available between the steam levels and the process sink curve. The experienced user of total site pinch analysis will recognize a clear signal to reduce the pressure in the MP steam mains.

This is implemented in Figure 38. The pressure of the MP steam mains has been reduced such that the amount of steam recovered from processes into the MP steam mains becomes equal to the amount of heat supplied. The MP steam mains is in balance and its load is maximized. Only a small pressure shift was found necessary. The results are dramatic. The amount of VHP steam which must be generated is reduced significantly.

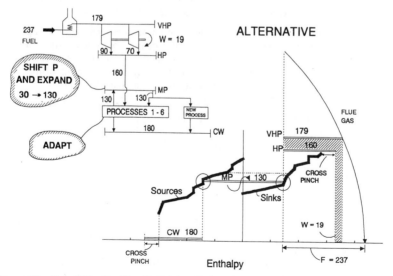

Figure 38—Total Site Profiles help identify an alternative strategy. If the pressure of the MP steam distribution level is reduced, investment in the infrastructure becomes minimal

Table 2

	Proposed (Figure 37)	Alternative (Figure 38)
Total Utilities Operating Cost	100%	81%
Capital cost implications	• Expand MP • New turbine • New boiler • Expand cooling	• Shift MP pressure & expand • Adapt processes to suit

Comparing Figure 37 (the conventional expansion plan) with Figure 38 (the alternative plan) we realize a 19% reduction in energy costs, see Table 2.

Surprisingly we also realize a significant reduction in capital costs[54]. Whereas the conventional expansion plan requires a new turbine, a new boiler and an expansion in the cooling system, the alternative plan requires no such investment in the infrastructure. Instead, there is moderate investment in the existing processes in order to enable the pressure of the MP mains to be downgraded. Table 2 offers a comparison of operating costs and capital cost factors. The design in Figure 38 represents a significant saving in investment *as well as* energy.

Please note that total site integration as described here does away with the age-old problem of steam pricing. Steam is strictly an intermediate parameter. All costs are explicitly stated in terms of fuel and power at the factory fence. The implications of any imaginable design or planning step are easily understood in terms of fuel and power at the factory fence. There is no need to assume a cost of steam.

Total site integration has been a major breakthrough in pinch analysis. Know-how has developed quickly, building on the considerable applications experience gained over many years in individual process integration. Successful industrial applications have already been reported in the literature[57,72,12,73].

Emissions Targeting

Emissions reduction and emissions targeting has a process and a total site dimension. The process dimension is referred to in Figure 39. Two process designs are shown, scheme A with a simple purge and scheme B with a more sophisticated separation/recycle concept reducing process waste. However, as is often the case, better separations involve additional energy. As indicated in Figure 39, the reduction in process emissions needs to be assessed relative to the increase in fuel related emissions. This dilemma is increasingly recognized both by designers and legislators[55,64]. There have been instances in the past decade where regulations requiring excessively low

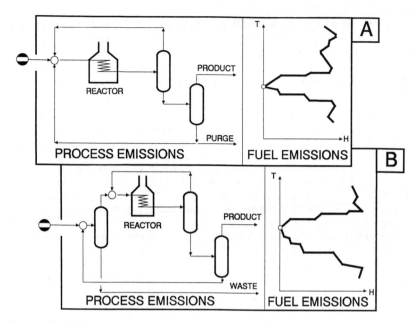

Figure 39—Different Process design options imply different levels of process emissions. However, they also imply different fuel consumption and therefore different levels of fuel related emissions. The overall emissions associated with any given process must therefore result from a correct trade-off between process emissions and fuel related emissions

ppm-limits on process emissions led to additional fuel consumption such that *overall* emissions deteriorated. It is clearly necessary to assess the overall picture. Pinch analysis targeting concepts have an obvious contribution to make in this context. They allow designers, planners, and legislators to come to a rational assessment of trade-offs between process related and fuel related emissions and to agree on achievable targets[69,71].

Over the last several years, emissions have taken on a total site dimension in many industrial situations. The so-called 'bubble concept' tends to guide planners and regulators in setting limits for emissions from either a company or a production site. Figure 40 relates to 'global emissions' (example CO_2) for a general site in the process industries. Total site pinch analysis as described in Figures 36, 37, and 38 easily addresses aspects such as production capacities, investments in the infrastructure, and investments in the processes. The analysis brings together these and other aspects in one coherent approach and gives targets for (1) central site combustion, (2) total site electric power import or export, and (3) de-centralized combustion on site. Global CO_2 emissions follow and can be targeted (e.g., as a function of investment) as shown in Figure 41.

Emissions targeting is specifically discussed elsewhere[54,64]. Pinch analysis can help predict both long and short term targets for specific processes and sites. The procedures involved are uncomplicated, accurate, and practical.

Discussion—Industrial Practice and Other Methods for Process Integration

Industrial practice
Pinch analysis found practical applications early in its history. The first publicly reported applications (from ICI, UK) averaged energy savings approaching 30% in processes previously thought optimized[37]. Applications reported by Union Carbide, USA, a few years later showed even better results mainly due to progress in the understanding of how to effect process changes[42]. BASF, Germany, reported completing over 150 projects and achieving site-wide energy savings of over 25% in retrofits in their main factory in Ludwigshafen[31]. They also reported significant environmental improvements. There have been many papers over the years from both operating companies and contractors reporting on the breadth of the technology, on applications, and on results achieved. In all, projects have been reported in over 30 countries.

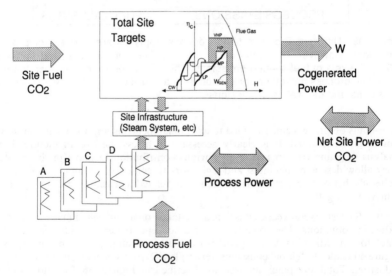

Figure 40—Total Site targeting allows us to bring together data describing the processes and data describing the utility systems in one coherent approach. Design changes, planning scenarios, capacity increases, etc., are immediately understood in terms of energy cost and fuel related emissions, regardless of fuel mix, power balance, etc. Notably, the approach is independent of any cost of steam. Steam is strictly an internal parameter

And yet, it is probably fair to say that pinch analysis remains relatively unknown in some countries and industries and perhaps controversial in others. Some believe the results reported have been exaggerated. Also, the early developments relating to heat exchanger network design remain better known than the later developments relating to overall processes and sites. Perhaps the early impact created too much attention too soon? For whatever reason, pinch analysis remains quite clearly under-utilized in industrial practice *relative to its potential*.

There are in existence several papers detailing industrial applications and impact[6,31,37,42,57,74]. For those who wish to get started, commercial software is available from Aspen[1], Linnhoff March[76], N.E.L.[25], SSI[26], and others. Training is available primarily from Linnhoff March[45] and UMIST[80].

Pinch analysis and other methodologies
So what is different about pinch analysis? Why have other methods for process integration found less application in practice?

One possible reason has already been referred to in the Introduction. Pinch analysis leaves the designer in control.

Mention should also be made of the description often given of pinch analysis as a *thermodynamic* method for process integration in contrast to other methods which

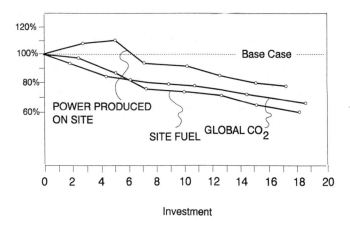

Figure 41—Targets are set as a function of investment for any specific site. Please note that in this example, based on a recent case study, power produced on site initially *increases* in spite of reduced site fuel consumption. Usually, we would expect co-generation potential to decrease as we burn less fuel. Experience has shown that better level matching between processes and utilities can lead to an increase in the amount of power produced per unit heat

are described as *mathematical*. This distinction deserves discussion. Many areas of engineering design combine physical insights with mathematical/numerical methods. For example, fluid dynamics give us the fundamentals, numerical methods allow us to apply these fundamentals in the design of anything other than simple flow systems. Physical insights, often based on thermodynamics, are the 'eye' of the engineer. Numerical methods are his or her 'muscle'. It is possible to argue that the maturity of any given area of engineering design is measured by the degree to which *both* thermodynamic (or physical) insights *and* numerical techniques are in use.

We recognize now that process integration is a relatively new area for engineering design methodology. Until recently, engineers developed conceptual process designs by experience and intuition without using methodology. In the opinion of the author, many of the so-called mathematical methods simply introduced 'muscle' before 'eye'. pinch analysis contributed more 'eye'. In this, pinch analysis was not alone. Exergy analysis, for example, aimed to improve insight a long time ago[63] and so did the heuristic approach recently summarised by J. Douglas[17]. However, process design methodology, as an area of interest, is still less developed than related areas (such as simulation or control) and there were few insights at the time. This is why the impact of pinch analysis was significant.

It is inevitable that the distinction between *mathematical* and *thermodynamic* will blur over time. Gundersen and Naess already discussed papers combining insights with numerical methods. Notable examples are papers from Westerberg[9], Floudas[19], Grossmann[60] and co-workers. Likewise, there is a growing body of powerful software supporting the more recent developments in pinch analysis discussed in this paper.

As use of methodology becomes the norm for conceptual process design, insights and numerical methods will come together more and more. Numerical methods will be used for increased power and speed of data handling, for the optimization of targets, for the automatic definition of key design patterns, etc. Pinch analysis will, over time, become less dominant. Instead, it will become part of, and perhaps be the basis of, a whole family of design methodologies, many of which will be software based, which will provide the process design engineer with a practical, mature and powerful working environment. This trend will be inevitable if the initial successes achieved with pinch analysis are to lead to general use of methodology in integrated process design.

Conclusions

Many conclusions could be drawn. Instead, the author would like to restrict himself to five key statements:

(1) Pinch analysis has become a mature tool. It works for practical problems. There is a track record of industrial applications which cannot be ignored.

(2) Pinch analysis has developed from a specialized tool for heat recovery to a broader based methodology for the conceptual design of process and energy systems including total sites.

(3) Today's major contribution is in analysis. Targets are analysed prior to design. This allows the designer to scope and screen broad design strategies.

(4) The scoping and screening considers overall economics (not just energy costs) and key feasibility features. Strategic alternatives are selected for large and seemingly complex design problems.

(5) Tomorrow's contribution lies in the provision of more insights, or fundamentals, and in their combination with numerical methods to give rise to innovative but practical design tools which will form a powerful working environment for conceptual process design.

In essence, pinch analysis reduces trial and error in conceptual process design. It does so through systematic analysis based on fundamental principles.

References

1. ADVENT, Pinch Technology Software, available from Aspen Technology, Inc., Ten Canal Park, Cambridge, Massachusetts 02141, USA.
2. Ahmad, S. and Smith, R., 1989, Targets and design for minimum number of shells in heat exchanger networks, *Chem Eng Res Des*, 67(5): 481–494.
3. Ahmad, S. and Hui, D. C. W., 1991, Heat recovery between areas of integrity, *Comp & Chem Eng*, 15(12): 809–832.
4. Ashton, G. J. *et al.*, Design and Operation of Energy Efficient Batch Processes. Report by Linnhoff March Limited, *Contract No. JOU-0043 (SMA)*, *JOULE Programme*, *Rational Use of Energy*, (The Commission of the European Communities).
5. Boland, D. and Linnhoff, B., 1979, The preliminary design of networks for heat exchange by systematic methods, *The Chemical Engineer*, 9–15, April.
6. Brown, K. J., 1989, Process Integration Initiative. A Review of the Process Integration Initiatives Funded under the Energy Efficiency R&D Programme. (Energy Technology Support Unit (ETSU), Harwell Laboratory, Oxon, UK), July.
7. Calandranis, J. and Stephanopoulos, G., 1986, Structural operability analysis of heat exchanger networks, *Chem Eng Res Des*, 64(5): 347–364.
8. Cerda, J., Westerberg, A. W., Mason, D. and Linnhoff, B., 1983, Minimum utility usage in constrained heat exchanger networks—A transportation problem, *Chem Eng Sci*, 38(3): 373–387.
9. Cerda, J. and Westerberg, A. W., 1983, Synthesizing heat exchanger networks having restricted stream/stream matches using transportation problem formulations, *Chem Eng Sci*, 38(10): 1723–1740.
10. Colberg, R. D. and Morari, M. and Townsend, D. W., 1989, A resilience target for heat exchanger network synthesis. *Comp & Chemical Eng*, 13(7): 821–837.
11. Colmenares, T. R. and Seider, W. D., 1987, Heat and power integration of chemical processes, *AIChE J.*, 33: 898–915
12. Davison, Robin, 1992, Pinch helps energy efficiency drive, *European Chemical News*, Special Report, 27 July.
13. Dhole, V. R. and Linnhoff, B., 1993, Distillation column targets, submitted and accepted by *Europ Symp on Computer Applications in Process Engineering ESCAPE-I, Elsinore, Denmark, May 24–28 1992*. (Extended version in *Comp & Chem Eng*, 17(5/6): 549–560.
14. Dhole, V. R. and Linnhoff, B., 1993, Total site targets for fuel, co-generation emissions, and cooling, paper presented at *ESCAPE-II Conference, Toulouse, France, October 1992*. Also in *Comp & Chem Eng*, 17(Suppl): s101–s109.
15. Dhole, V. R. and Linnhoff, B., 1993, Overall design of subambient plants, paper presented at *ESCAPE-III Conference, Graz, Austria, July*. Also in *Comp & Chem Eng*, 1994, 18(Suppl): S105–S111.

16. Dhole, V. R. and Zheng, J. P., 1993, Applying combined pinch and exergy analysis to closed cycle gas turbine system design, paper presented at *ASME Cogen Turbo Power, The 7th Congress and Exposition, Bournemouth*, UK. *September*. Also to appear in *ASME J Eng*, 1994, as Gas turbines and power.

17. Douglas, J. M., 1988, *Conceptual Design of Chemical Processes*, (McGraw-Hill, New York, N.Y.).

18. El-Halwagi, M. M. and Manousiouthakis, V., 1989, Synthesis of mass-exchange networks, *AIChE J*, 8: 1233–1244.

19. Floudas, C. A., Ciric, A. R. and Grossmann, I. E., 1986, Automatic synthesis of optimum heat exchanger network configurations, *AIChE J*, 32(2): 276–290.

20. Floudas, C. A. and Grossmann, I. E., 1987, Synthesis of flexible heat exchanger networks for multiperiod operation, *Comp & Chem Eng*, 11(2): 123–142.

21. Fonyo, Z., 1974, Thermodynamic analysis of rectification 1. Reversible model of rectification, *Int Chem Eng*, 14, 18–27.

22. Gremouti, I. D., 1991, Integration of batch processes for energy savings and debottlenecking, *MSc Thesis*, (University of Manchester (UMIST), UK).

23. Gundersen, T. and Naess, L., 1988, The synthesis of cost optimal heat exchanger networks —An industrial review of the state of the art, *Comp & Chem Eng*, 12(6): 503–530.

24. Hall, S. G., Ahmad, S. and Smith, R., 1990, Capital cost targets for heat exchanger networks comprising mixed materials of construction, pressure ratings and exchanger types, *Comp & Chem Eng*, 14(3): 319–335.

25. HEATNET, Pinch Analysis Software, available from N.E.L., East Kilbride, Glasgow, G75 0QU.

26. HEXTRAN, Software for Heat Exchanger Network Design and Simulation, available from Simulation Sciences, Brea, CA, USA.

27. Huang, F. and Elshout, R. V., 1976, Optimizing the heat recovery of crude units, *Chem Eng Prog*, 72(7): 68–74.

28. Jones, P. S., 1991, Targeting and design for heat exchanger networks under multiple base case operation, *PhD Thesis*, (University of Manchester (UMIST), UK).

29. Kaibel, G., 1987, *PhD Thesis*, (Technical University, Munich, Germany).

30. Kemp, I. C. and Deakin, A. W., 1989, The cascade analysis for energy and process integration of batch processes, *Chem Eng Res Des*, 67: 495–525.

31. Korner, H., 1988, Optimal use of energy in the chemical industry, *Chem Ing Tech*, 60(7): 511–518.

32. Kotjabasakis, E. and Linnhoff, B., 1986, Sensitivity tables for the design of flexible processes (1)—How much contingency in heat exchanger networks is cost-effective, *Chem Eng Res Des*, 64: 197–211.

33. Kotjabasakis, E. and Linnhoff, B., 1987, Better system design reduces heat-exchanger fouling costs, *Oil and Gas Journal*, 49–56, Sept.

34. Linnhoff, B. and Flower, J. R., 1978, Synthesis of heat exchanger networks: Part I: Systematic generation of energy optimal networks, *AIChE J*, 24(4): 633–642. Part II: Evolutionary generation of networks with various criteria of optimality, *AIChE J*, 24(4): 642–654.

35. Linnhoff, B., Mason, D. R. and Wardle, I., 1979, Understanding heat exchanger networks, *Comp & Chem Eng*, 3: 295–302.

36. Linnhoff, B. and Turner, J. A., 1980, Simple concepts in process synthesis give energy savings and elegant designs, *The Chemical Engineer*, 742–746, December.

37. Linnhoff, B. and Turner, J. A., 1981, Heat-recovery networks: new insights yield big savings, *Chem Eng*, 56–70, Nov. 2.

38. Linnhoff, B., Townsend, D. W., Boland, D., Hewitt, G. F., Thomas, B. E. A., Guy, A. R. and Marsland, R. H., 1982, *User Guide on Process Integration for the Efficient Use of Energy*, (IChemE, Rugby, UK). References to specific sections are given in the text, as appropriate.

39. Linnhoff, B. and Hindmarsh, E., 1983, The pinch design method of heat exchanger networks, *Chem Eng Sci*, 38(5): 745–763.
40. Linnhoff, B., Dunford, H. and Smith, R., 1983, Heat integration of distillation columns into overall processes, *Chem Eng Sci*, 38(8): 1175–1188.
41. Linnhoff, B. and Parker, S., 1984, Heat exchanger networks with process modifications, *IChemE Annual Research Meeting, Bath*, UK. *April*.
42. Linnhoff, B. and Vredeveld, D. R., 1984, Pinch technology has come of age, *Chem Eng Prog*, 33–40, July.
43. Linnhoff, B. and Witherell, W. D., 1986, Pinch technology guides retrofit, *Oil & Gas Journal*, 54–65, 7 April.
44. Linnhoff, B. and Kotjabasakis, E., 1986, Process optimization: downstream paths for operable process design, *Chem Eng Prog*, 23–28, May.
45. Linnhoff, B. *et al.*, 1986, *Pinch Analysis* Foundation Training Course, © 1986, (Linnhoff March Ltd., Tabley Court, Moss Lane, Over Tabley, Knutsford, WA16 0PL, UK).
46. Linnhoff, B. and Lenz, W., 1987, Thermal integration and process optimization, *Chem Eng Tech*, 59: 851–857.
47. Linnhoff, B. and Polley, G., 1988, Stepping beyond the pinch, *The Chemical Engineer*, 25–32, February.
48. Linnhoff, B. and de Leur, J., 1988, Appropriate placement of furnaces in the integrated process paper presented at *IChemE Symposium 'Understanding Process Integration II' 22–23 March, UMIST, Manchester*. UK.
49. Linnhoff, B., Polley, G. T. and Sahdev, V., 1988, General process improvements through pinch technology, *Chem Eng Prog*, 51–58, June.
50. Linnhoff, B., Pinch technology for the synthesis of optimal heat and power systems, *Trans ASME, J of Energy Resources Technol*, 111(3): 137–147.
51. Linnhoff, B. and Alanis, F. J., 1989, A systems' approach based on pinch technology to commercial power station design, paper presented at *ASME Winter Annual Meeting, San Francisco, Ca. USA* 10–15 December.
52. Linnhoff, B. and Ahmad, S., 1990, Cost optimum heat exchanger networks, Part 1: minimum energy and capital using simple models for capital cost, *Comp & Chem Eng*, 14(7): 729–750. Ahmad, S., Linnhoff, B. and Smith, R., Part 2: Targets and design for detailed capital cost models, *Comp & Chem Eng*, 14(7): 751–767.
53. Linnhoff, B. and Dhole, V. R., 1992, Shaftwork targets for low temperature process design, *Chem Eng Sci*, 47(8): 2081–2091.
54. Linnhoff, B. and Dhole, V. R., 1993, Targeting for CO_2 emissions for total sites, *Chem Eng Technol*, 16: 252–259
55. Linnhoff March Inc., Leesburg, Va., 1991, Expanded pinch analysis procedure for pollution prevention at Amoco's Yorktown, Va. refinery. (Report prepared for Amoco Corp., Chicago, Ill. and United States Environmental Protection Agency, Washington, DC, 6 June publicly available).
56. Morgan, S., 1992, Use process integration to improve process designs and the design process, *Chem Eng Prog*, 62–68, September.
57. Natori, Y., 1992, Managing the implementation of pinch technology in a large company, paper presented at the *IEA Workshop on Process Integration, Gothenburg, Sweden, January 28–29*.
58. Obeng, E. D. A. and Ashton, G. J., 1988, On pinch technology based procedures for the design of batch processes, *Chem Eng Res Des*, 66: 255–259.
59. O'Young, L., 1989, Constrained heat exchanger networks: targeting and design, *PhD Thesis*, (University of Manchester (UMIST), UK).
60. Papoulias, S. A. and Grossmann, I. E., 1983, A structural optimization approach in process synthesis—II, Heat recovery networks, *Comp & Chem Eng*, 7(6): 707–721.

61. Parker, S. J., 1989, Supertargeting for multiple utilities, *PhD Thesis*, (University of Manchester (UMIST), UK).
62. Polley, G. T., Panjeh Shahi, M. H. and Jegede, F. O., 1990, Pressure drop considerations in the retrofit of heat exchanger networks, *Trans IChemE*, 68, (Part A): 211–220.
63. Rant, Z., 1956, Exergie, ein neues Wort für technische Arbeitsfähigkeit, *Forschungs-Ingenieur-Wesen*, 22: 36.
64. Rossiter, A. P., Spriggs, H. D. and Klee, H., 1993, Apply process integration to waste minimization, *Chem Eng Prog*, 30–36.
65. Saboo, A. K., Morari, M. and Woodcock, D. C., 1985, Design of resilient processing plants: VIII. A resilience index for heat exchanger networks, *Chem Eng Sci*, 40(8): 1553–1565.
66. Saboo, A. K., Morari, M. and Colberg, R. D., 1986, RESHEX—An interactive software package for the synthesis and analysis of resilient heat exchanger networks, Part I: Program description and application, *Comp & Chem Eng*, 10(6): 577–589. Part II: Discussion of area targeting and network synthesis algorithms, *Comp & Chem Eng*, 10(6): 591–599.
67. Shokoya, C. G., 1992, Retrofit of heat exchanger networks for debottlenecking and energy savings, *PhD Thesis*, (University of Manchester (UMIST), UK).
68. Smith, R. and Linnhoff, B., 1988, The design of separators in the context of overall processes, *Chem Eng Res Des*, 66(3): 195–228.
69. Smith, R., Petela, E. A. and Spriggs, H. D., 1990, Minimization of environmental emissions through improved process integration, *Heat Recovery Systems & CHP*, 10(4): 329–339.
70. Smith, R. and Jones, P. S., 1990, The optimal design of integrated evaporation systems, *Heat Recovery Systems & CHP*, 10(4): 341–368.
71. Smith, R. and Petela, E. A., 1991–1992, Series of five papers on Waste Minimisation and Process Design: 1 The Problem, *The Chemical Engineer*, 24–25, October, 1991, 2: Reactors, *ibid* 17–23, December 1991, 3: Separation and Recycle Systems, *ibid* 24–28, February 1992, 4: Process Operations, *ibid* 21–23, April 1992, 5: Utility Waste, *ibid* 32–35, July 1992.
72. Snoek, J. and Tjoe, T. N., 1992, Process integration experience in a large company. Paper presented at the *IEA Workshop on Process Integration, Gothenburg, Sweden, January 28–29*.
73. Spriggs, D. H., 1992, Experiences in total site studies, paper presented at the *IEA Workshop on Process Integration, Gothenburg, Sweden, January 28–29*.
74. Steinmetz, F. J. and Chaney, M. O., 1985, Total plant process energy integration, *Chem Eng Prog*, 81: 27–32, July.
75. Steinmeyer, D., 1992, Save energy, without entropy, *Hydro Carbon Processing*, 71: 55–95, October.
76. SUPERTARGET, Pinch Analysis Software, available from Linnhoff March Ltd., Tabley Court, Moss Lane, Over Tabley, Knutsford, WA16 0PL, UK.
77. Tjoe, T. N. and Linnhoff, B., 1986, Using pinch technology for process retrofit, *Chem Eng*, 47–60, April 28.
78. Townsend, D. W. and Linnhoff, B., 1982, Designing total energy systems by systematic methods, *The Chemical Engineer*, 91–97, March.
79. Townsend, D. W. and Linnhoff, B., 1983, Heat and power networks in process design, Part I: Criteria for placement of heat engines and heat pumps in process networks, *AIChE J*, 29(5): 742–748. Part II: Design procedure for equipment selection and process matching, *AIChE J*, 29(5): 748–771.
80. UMIST, MSc Course in Process Integration. Taught since 1984 by the Department for Process Integration, UMIST, PO Box 88, Manchester, UK.
81. Wang, Y. P. and Smith, R., 1993, Wastewater minimization, submitted to *Chem Eng Sci*.

2. Network Integration

2.1 The Role of Thermodynamics in Process Design

2.2 Heat Exchanger Networks

2.2.1. An appreciation of the integration techniques

2.2.2. Energy targets
2.2.2.1. For two streams only
2.2.2.2. For many streams—"Composite Curves"
2.2.2.3. A targeting procedure—the "Problem Table"

2.2.3. Simple design for maximum energy recovery
2.2.3.1. The significance of the "Pinch"
2.2.3.2. Network representation
2.2.3.3. Design for best energy recovery
2.2.3.4. A word about design strategy

2.2.4. Trading off energy against capital
2.2.4.1. The minimum number of units
2.2.4.2. Targeting for the minimum number
2.2.4.3. Trading off units and energy
2.2.4.4. The role of ΔT_{min}—trading off area and energy

2.2.5. Multiple utilities
2.2.5.1. Understanding process sources and sinks—the "Grand Composite Curve"
2.2.5.2. Designing for many utilities
2.2.5.3. Multiple utilities design example

2.2.6. Applying the principles—some further aspects
2.2.6.1. Is there a trade-off?—"threshold problems"
2.2.6.2. Stream-splitting
2.2.6.3. Design away from the "Pinch"
2.2.6.4. Constraints
2.2.6.5. Revamp studies

2.2.7. Data extraction (or, are you solving the right problem?)
2.2.7.1. Data accuracy
2.2.7.2. Choosing streams

2.1.   The Role of Thermodynamics in Process Design

Most of us involved in engineering design have somewhat unhappy memories thinking back to thermodynamics in college days. Either we did not understand, gave up hope that we ever would, and dread remembering the horror that struck on examination

8

day. Alternatively, we were amongst the chosen few whose photographic memory would allow us to reiterate the definitions of entropy, Gibbs-free energy and all those differential equations faultlessly, but without understanding. Afterwards, we could never help asking ourselves: what is it all for? What do I *do* with it? In the best of cases, thermodynamics seemed to be a fascinating science without a real application.

As mentioned in the Introduction, much of the work on which this Guide is based is straight-forward thermodynamics (Linnhoff, 1979). However, the approch is entirely non-mathematical. The starting premise is that (classical) thermodynamics itself may be a thoroughly developed subject, but that the interpretation of thermodynamics in the context of practical design needs much development. What follows are definitions of "inevitable" *versus* "avoidable" thermodynamic losses, of "practical" *versus* "ideal" performance targets and a good number of individual concepts and insights of great practicality. Direct spin-offs of this work are the network integration techniques described in this Guide and the main feature they share with their origin is the idea of practical performance targets.

Practical thermodynamic performance targets
Prior to any actual design being carried out, the techniques to be presented focus on setting practical performance targets, both for the achievable energy performance of a system, and for the achievable number of capital items (heat transfer "units"). To

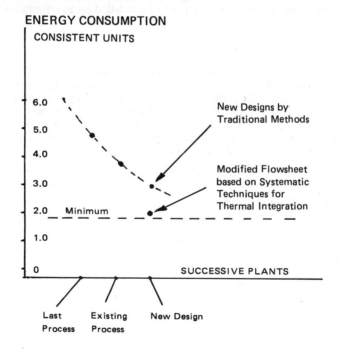

Figure 2.1—Energy consumption for traditional designs and alternative flowsheet (second case study)

9

give an example, it is possible to predict with little effort that for the system in Figure 1.1 the minimum number of heat transfer "units" is four and the minimum steam requirement is 1068 units of heat. These targets can either give the designer confidence or they can act as a stimulus. If a design achieves the targeted performance as in Figure 1.1(b), the designer can be confident that further improvement is not possible. If a design does not achieve the targeted performance as in Figure 1.1(a) the designer is stimulated to improve it.

The effect which such a stimulus can have in a practical situation is described in Figure 2.1 (Boland and Linnhoff, 1979). The Figure shows the improvement in energy consumption which was achieved by successive designs for a given product. The successive designs lie on a "learning curve". However, application of "targets" as described in this Guide revealed suddenly that the ultimate performance, given correct integration, would lie quite a bit further down the "learning curve". This information acted as an enormous stimulus to the design team. Within a short period they produced a flowsheet virtually "hitting" the ultimate practical target.

It is interesting to note that the concept of "learning curves" as shown in Figure 2.1 is, in essence, a targeting concept, too. The performance of new designs is stimulated by "learning curves" to show at least some improvement over the old ones. The difference between "learning curves" and the targeting procedures presented in this Guide is that "learning curves" set targets based on an extrapolation of the past, while the techniques presented in this Guide set targets based on an objective analysis.

Can thermodynamics help save capital costs?
In most people's minds, thermodynamics is associated with energy costs and thermodynamic arguments are only practical if capital costs are low. Consider, for example, Figure 2.2. At the top a heat exchanger network is shown that would seem appropriate to most when energy is cheap and capital expensive. There is no process heat recovery—only utility usage. At the bottom a network is shown which would seem appropriate to most when energy is expensive. There is as much process heat recovery as is possible in preference to utility usage. The implicit assumption is that heat recovery (instead of utility use) saves energy but costs capital.

Consider now Figure 2.3. Based on a uniform heat transfer coefficient and sensible steam and cooling water temperatures, the total surface area for both designs has been evaluated. To our surprise, the "network for minimum capital cost" turns out to have the higher total surface area!

From this example we realise that in networks there are two basic thermodynamic effects influencing capital costs. One is the effect of driving forces and the other is the effect of heat loads; see Figure 2.4. Evidently, as we go to tighter designs (*i.e.* to reduce driving forces) we need less utility and the overall heat load decreases. Captial cost then *increases* with reduced driving forces (we all know that) but *decreases* with reduced heat load (we rarely consider this point). In the example in Figures 2.2 and 2.3, the design without process heat recovery handles twice as much heat as is necessary. As a result, capital costs are increased even though the driving forces are large!

The example in Figures 2.2 and 2.3 may appear to be, and indeed is, contrived.

10

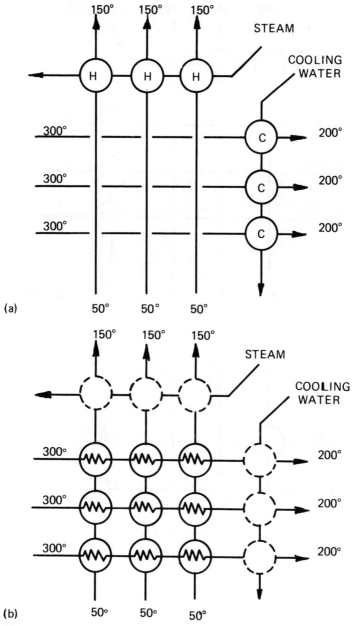

Figure 2.2—Networks for (a) minimum capital cost and (b) for minimum energy cost

11

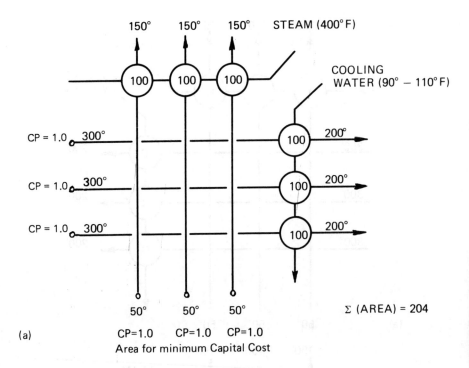

(a) Area for minimum Capital Cost

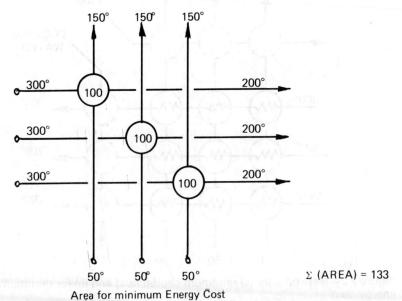

(b) Area for minimum Energy Cost

Figure 2.3—Area for (a) minimum capital cost and (b) minimum energy cost

12

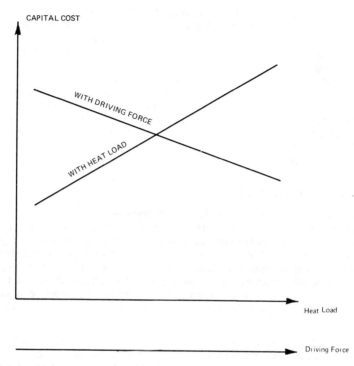

Figure 2.4—Effect of driving force and heat load on capital cost

However, its purpose is to clarify the argument portrayed in Figure 2.4. This argument is certainly general and valid. It helps to explain the frequent capital savings observed in practical case studies as listed in Table 1.1.

Thermodynamics identifies options

There is another issue that is worth taking up with those that believe that thermodynamic arguments are only relevant if capital costs are low. It is probably fair to say that this notion is based on a somewhat narrow appreciation of thermodynamics: thermodynamics is used to analyse driving forces in a design and the existing driving forces are then reduced so that energy is saved while capital is spent. Why not use the analysis of driving forces not to reduce them but to distribute them differently? This can help to clarify options in design, say, for better operability and/or for lower capital costs at a constant level of energy recovery. Thus, more generally than just saving energy, thermodynamics can help by identifying *options*.

The reader will soon recognise that the techniques described in this Guide do not attempt to save energy as an end in itself. Rather, they attempt to identify options. These options can then point towards possible energy savings, capital savings, or preferred integration alternatives in the interests of given constraints in plant lay-out, control, safety, *etc.*

13

Further, the options can be used to study trade-offs between conflicting criteria. To demonstrate the power of the techniques in this respect it is worth noting that they allow the user not only to study well-known trade-offs such as energy recovery *versus* surface area in heat exchangers. Beyond this, they also highlight trade-offs that were hitherto improperly understood. The main example here is the trade-off between energy recovery and the number of heat transfer "units". The number of separate heat transfer units (heaters, coolers and exchangers) necessary for full heat recovery is greater than for reduced heat recovery, independent of total surface area. The techniques explain the principles of this phenomenon and offer a procedure for dealing with the trade-off in practical studies. Generally speaking, the reader will recognise when reading the Guide that thermodynamic principles can be practical, whether or not energy is expensive.

2.2 Heat Exchanger Networks

2.2.1 An appreciation of the integration techniques

In the previous section of the Guide the claim was made that a proper analysis of thermodynamics can lead to significant improvements in process plant energy performance. Not only that, it was claimed that an appropriate thermodynamic analysis can also lead to identification of preferred options in terms of many other design objectives, for example minimum capital cost and operability. These are substantial claims, and it is the purpose of this sub-section of the Guide to give the reader confidence that these claims are justified. Hopefully, this should motivate him or her to carry on reading into the main technical sections!

It is possible to identify heat recovery as a separate and distinct task in process design. Figure 2.5 shows a more detailed flowsheet for the front end of the specialty chemicals process which was discussed in the Introduction. In terms of unit operations, three distinct tasks are being performed, namely reaction, separation and heat exchange.

The design of the reactor is dictated by yield and conversion considerations, and that of the separator by the need to flash off as much unreacted feed as possible. If the operating conditions of these units are accepted, then the design problem that remains is to get the optimum economic performance out of the *system* of heat exchangers, heaters and coolers. The design of the heat exchange system or "network" as it stands in Figure 2.5 may not be the best and so it is necessary to go back to the underlying data that define the problem.

The basic elements of the heat recovery problem are shown in Figure 2.6. All the exchangers, heaters and coolers have been "stripped out" of the flowsheet and what remains therefore is the definition of the various heating and cooling tasks. Thus one stream, the reactor product, requires cooling from reactor exit temperature to separator temperature. Three streams require heating, these being reactor feed (from fresh feed storage temperature to reactor inlet temperature), recycle (from recycle temperature to reactor inlet temperature), and the "front end" product (from separator temperature to the temperature needed for downstream processing). Therefore the problem data comprise a set of four streams, one requiring cooling and three requiring heating, whose endpoint temperatures are known and whose total enthalpy changes are known (from the flowsheet mass balance and physical proper-

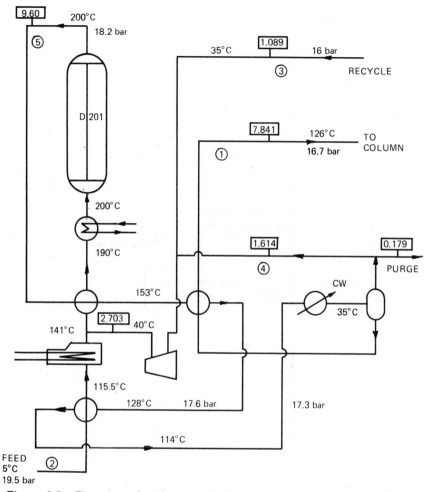

Figure 2.5—Flowsheet for "front end" of specialty chemicals process

ties). The design task is to find the best network of exchangers, heaters and coolers, that handle these four streams at minimum operating and annualised capital cost, consistent with other design objectives such as operability.

The total cost of such a network tends to be dominated by the size of the steam and cooling water demands and by the number of capital items in the network (*i.e.* total number of heaters, coolers and exchangers). For a "first go" at the design, therefore, we can modify the objective to producing a design which consumes minimum quantities of utilities and employs the smallest possible number of "units" (*i.e.* heaters, coolers, exchangers). At first sight, in a problem comprising only four process streams, this may seem an easy task. The reader might therefore like to try solving a simplified example problem comprising four process streams (*i.e.* similar to the pro-

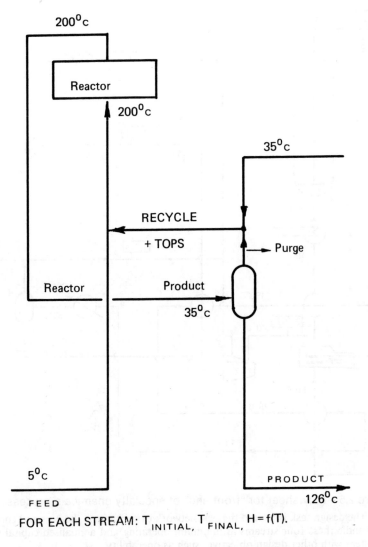

FOR EACH STREAM: $T_{INITIAL}$, T_{FINAL}, $H = f(T)$.

Figure 2.6—Specialty chemicals process—heat exchange duties

cess example of Figure 2.5), the data for which are given in Table 2.1. Note that in this example we have two "hot" streams (*i.e.* streams that have to be cooled) rather than one, and only two "cold" streams (*i.e.* streams that have to be heated). Interchangers may not have a temperature difference between the hot and cold streams of less than 10°C. Steam which is sufficiently hot and cooling water which is sufficiently cold for any required heating and cooling duty is available. After trying this example, the

Table 2.1

Process stream number and type	Heat capacity flowrate (kW/°C)	Supply temperature (°C)	Target temperature (°C)	Stream heat duty (kW)
(1) *cold*	2.0	20	135	2.0 × (135 − 20) = 230
(2) *hot*	3.0	170	60	3.0 × (60 − 170) = − 330
(3) *cold*	4.0	80	140	4.0 × (140 − 80) = 240
(4) *hot*	1.5	150	30	1.5 × (30 − 150) = − 180

$\Delta T_{min} = 10°C$

Utilities: steam at 200°C
cooling water at 15°C

reader will probably agree that it is not a trivial task. However, if he or she had known *before starting* that the best possible energy performance for this problem, observing the 10°C constraint, is 20 units of heating and 60 units of cooling, then this would have provided a big stimulus to improving on first attempts. Furthermore, prior understanding that to solve the problem with minimum energy requires a minimum of six units is also very useful.

The capability to set performance targets prior to design is an immensely powerful one, since it stimulates the designer to keep on trying until he or she achieves the best possible (Linnhoff and Turner, 1981). In the remaining part of this section we set out to demonstrate that such target-setting is indeed possible.

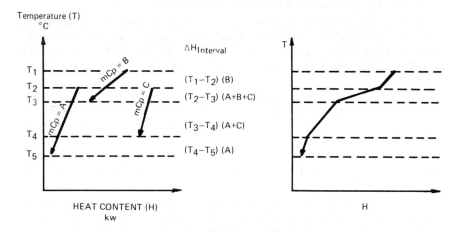

Figure 2.7—Construction of "Composite Curves"

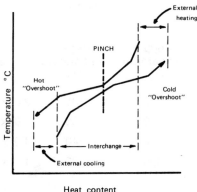

Figure 2.8—Prediction of energy targets using Composite Curves

The hot and cold streams in a process can be represented on a temperature-heat content (enthalpy) graph once their input and output temperatures (or "supply" and "target" temperatures) and their flowrates and physical properties are known.

Starting from the individual streams it is possible to construct one "composite curve" of all hot streams in the process and another of all cold streams, by simple addition of heat contents over the temperature ranges in the problem.

This is illustrated in Figure 2.7 for a number of hot streams with heat capacities A, B and C being cooled through the temperature levels indicated. The procedure is described in more detail in the next sub-section.

The result for a set of hot and cold streams is a plot of two composite curves as shown in Figure 2.8. The overlap between the composite curves represents the maximum amount of heat recovery possible within the process.

The "over-shoot" of the hot composite represents the minimum amount of external cooling required and the "over-shoot" of the cold composite represents the minimum amount of external heating (Hohmann, 1971). Because of the "kinked" nature of the curves, they approach most closely at one point. This is the "pinch" (Linnhoff et al, 1979).

In Figure 2.9 (a) the system is shown separated at the pinch. In the section above the pinch, the composite hot gives all its heat to the composite cold with only residual heating required. The system is therefore a heat sink. Heat goes in from hot utility, but no heat goes out.

Conversely, below the pinch the system is a heat source. Heat goes out to cold utility but no heat goes in. Hence in a design that achieves the utility targets the heat flow across the pinch is zero.

However, Figure 2.9 (b) shows the case where the minimum utility targets are not met. External heating is in excess (by α) of the minimum possible. By heat balance around the heat source and the heat sink, there must then be a heat flow α across the pinch and an excess external cooling requirement α.

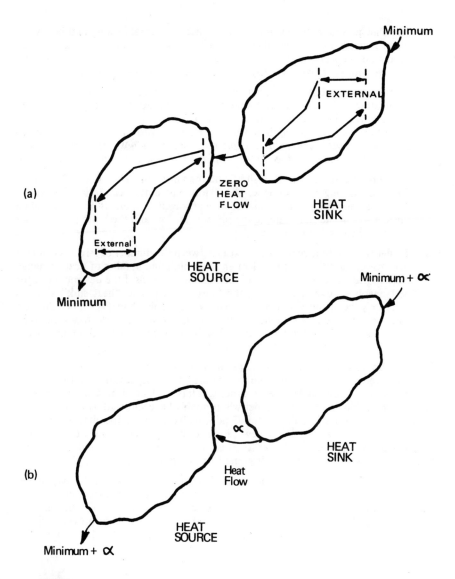

Figure 2.9—The source/sink characteristic of process heat exchange systems

This insight gives us five concepts which are as effective as they are simple:

● **Targets**: Once the composite curves are known, we know exactly how much external heating is required. *Near-optimal processes are confirmed as such and non-optimal processes are identified with great speed and confidence.*

- **The Pinch**: Above the pinch the process needs external heating and below the pinch it needs external cooling. This tells us where to place furnaces, steam heaters, coolers *etc.* It also tells us what site steam services should be used and how we should recover heat from the exhaust of steam and gas turbines.

- **More in, more out**: An inefficient process requires more than the minimum external heating and therefore more than the minimum external cooling; see Figure 2.9 (b). We coin the catch phrase "more in, more out" and note that for every unit of excess external heat in a process we have to provide heat transfer equipment twice. *This insight helps us explain why it is often possible to improve both energy and capital cost. (See Table 1.1 in the Introduction section of the Guide).*

- **Freedom of choice**: The "heat sink" and the "heat source" in Figure 2.9 (b) are separate. As long as the designer obeys this constraint he can follow his heart's delight in choosing plant-layouts, control arrangements, *etc.* If he has to violate this constraint, he can evaluate the pinch heat flow and therefore predict what overall penalties will be involved.

- **Tradeoffs**: A simple relationship exists between the number of streams (process streams plus utilities) in a problem and the minimum number of heat exchange "units" (*i.e.* heaters, coolers, and interchangers). Thus if the designer goes for best energy recovery, designing the "heat source" and "heat sink" sections separately, he or she will incur the need for more units than if the pinch division had been ignored. *Hence a new type of trade-off has been identified, between energy recovery and number of units.* This insight adds to the traditional concept of a trade-off between energy and surface area.

Thus, we do not need "black-box" computing power, but we can blend basic rules such as "keep the sink and the source separate" with the designer's intuition and experience of the individual process technology. It is this blend which ultimately gives better designs. However, as the reader will probably have gathered by now, the heat integration techniques described in this Guide involve many significant new concepts, upon which is built a completely new design technology. This means that concentrated effort will be required to get to grips with the material presented. There's no getting round it:—from now on it's going to be hard work!

In the sections which follow under "Heat Exchanger Networks", information will not be presented in the order in which it is ultimately applied in real engineering studies. Rather, it is presented in a sequence which will enhance the reader's understanding. Essentially, the philosophy of the description is to present a quick "run through" of some of the basics leading to network design. Detail is then filled in later.

2.2.2. *Energy targets*

2.2.2.1. *Two stream heat exchange*—The temperature-heat content (temperature-enthalpy) diagram can be used to represent the thermal characteristics of process streams, as illustrated in Figure 2.10. Differential heat flow dQ, when added to a process stream, will increase its enthalpy (H) by $CP.dT$, where:

CP = "heat capacity flowrate" (kW/K) = mass flow (kg/sec) $\times$ specific heat (kJ/kgK)
dT = differential temperature change

Hence, with *CP* assumed constant, for a stream requiring heating ("cold" stream) from a "supply temperature" (T_s) to a "target temperature" (T_T), the total heat added will be equal to the stream enthalpy change, *i.e.*

$$Q = \int_{T_S}^{T_T} CP.\mathrm{d}T = CP(T_T - T_s) = \Delta H$$

and the slope of the line representing the stream is

$$\frac{\mathrm{d}T}{\mathrm{d}Q} = \frac{1}{CP}$$

The T/H diagram can be used to represent heat exchange, because of a very useful feature. Namely, since we are only interested in enthalpy *changes* of streams, a given stream can be plotted anywhere on the enthalpy axis. Provided it has the same slope and runs between the same supply and target temperatures, then wherever it is drawn on the H-axis, it represents the same stream.

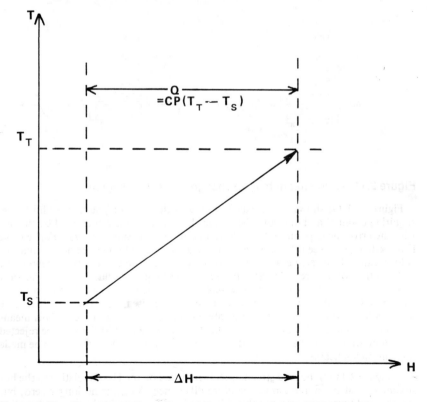

Figure 2.10—Representation of process streams in the T/H diagram

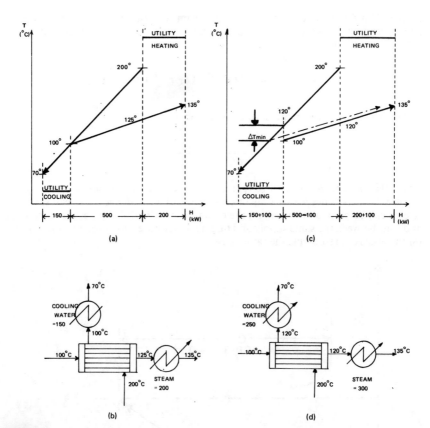

Figure 2.11—Two-stream heat exchange in the T/H diagram

Figure 2.11 (a) shows two streams plotted in the T/H diagram, one "hot" (*i.e.* requiring cooling) and one cold. Note that the hot stream is represented by the line with the arrow head pointing to the left, and the cold stream *vice versa*. For feasible heat exchange between the two, the hot stream must at all points be hotter than the cold stream. However, because of the relative temperatures of the two streams, the construction of Figure 2.11 (a) represents a limiting case illustrated by the flow diagram in Figure 2.11 (b). Heat exchange between the hot stream running counter-current to the cold stream could not be increased because the temperature difference between hot and cold streams at the cold end of the exchanger is zero. This means that, in this example; heat available in the hot stream below 100°C has to be rejected to cooling water, and the balance of heat required by the cold stream has to be made up from steam heating.

In Figure 2.11 (c) the cold stream is shown shifted on the H-axis relative to the hot stream so that the minimum temperature difference, ΔT_{min}, is no longer zero, but positive and finite. The effect of this shift is to increase the utility heating and cooling

22

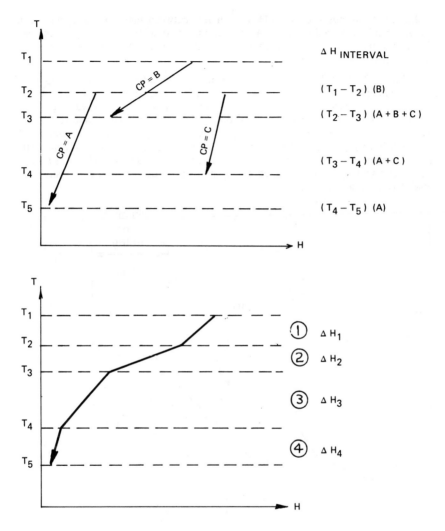

Figure 2.12—Construction of Composite Curves

by equal amounts and reduce the load on the exchanger by the same amount. The arrangement, which is now practical because the $\Delta T_{\min}$ is non-zero, is shown in flow diagram form in Figure 2.11 (d). Clearly, further shifting implies larger $\Delta T_{\min}$ values and larger utility consumptions.

From this analysis, two basic facts emerge. Firstly, there is a correlation between the value of $\Delta T_{\min}$ in the exchanger and the total utility load on the system. Secondly, if the hot utility load is increased by any value α, the cold utility is increased by α as well. *More in, more out!*

2.2.2.2. Composite curves—Heat exchange between many hot and many cold streams can be analysed similarly. A single composite of all hot and a single composite of all cold streams can be produced in the T/H diagram, and handled in just the same way as the two-stream problem.

In Figure 2.12 (a) three hot streams are plotted separately, with their supply and target temperatures defining a series of "interval" temperatures T_1 to T_5. Between T_1 and T_2, only stream B exists, and so the heat available in this interval is given by $CP_B(T_1 - T_2)$. However between T_2 and T_3, all three streams exist and so the heat available in this interval is $(CP_A + CP_B + CP_C)(T_2 - T_3)$. A series of values of ΔH for each interval can be obtained in this way, and the result re-plotted against the interval temperatures as shown in Figure 2.12 (b). The resulting T/H plot is a single curve representing all the hot streams. A similar procedure gives a composite of all cold streams in a problem.

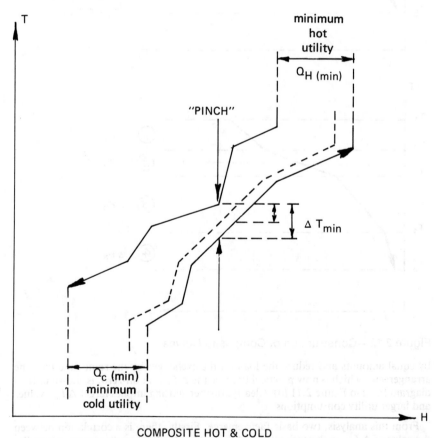

COMPOSITE HOT & COLD

Figure 2.13—Energy targets and "the Pinch" with Composite Curves

24

Figure 2.13 shows a typical pair of composite curves. Shifting of the curves leads to behaviour similar to that shown by the two-stream problem. Now, though, the "kinked" nature of the composites means that ΔT_{min} can occur anywhere in the interchange region and not just at one end. *For a given value of ΔT_{min}, the utility quantities predicted are the minima required to solve the heat recovery problem.* Note that although there are many streams in the problem, in general ΔT_{min} occurs at only one point, termed the "pinch". This means that it is possible to design a network which uses the minimum utility requirements, where *only the heat exchangers at the pinch* need to operate at ΔT values down to ΔT_{min}. Producing such a design will be described later.

2.2.2.3. A targeting procedure—the "Problem Table"—In principle, the "composite curves" described in the previous sub-section could be used for obtaining energy targets at given values of ΔT_{min}. However, it would require a "graph paper and scissors" approach (for sliding the graphs relative to one another) which would be messy and imprecise. An algorithm for setting the targets algebraically has been published by Linnhoff and Flower (1978), known as the "Problem Table" method. Their method is described here.

In the description of the construction of composite curves (see Figures 2.12 (a) and (b)), it was shown how enthalpy balance intervals were set up based on stream supply and target temperatures. The same can be done for hot and cold streams together, to allow for the maximum possible amount of heat exchange within each temperature interval. The only modification needed is to ensure that within any interval, hot streams and cold streams are at least ΔT_{min} apart. This is done in the way shown in Figure 2.14 (a) for the problem data from Table 2.1. Streams are shown in a schematic representation with a vertical temperature scale. Temperature interval boundaries are superimposed. The interval boundary temperatures are set at $1/2\Delta T_{min}$ (5°C in this example) *below* hot stream temperatures and $1/2\Delta T_{min}$ *above* cold stream temperatures. So for example in interval number 2 in Figure 2.14 (a), streams 2 and 4 (the hot streams) run from 150°C to 145°C, and stream 3 (the cold stream) from

Table 2.1 (repeat)

Stream No. And Type	CP (kw/°C)	$T_s(°C)$	$T_t(°C)$
(1) Cold	2	20°	135°
(2) Hot	3	170°	60°
(3) Cold	4	80°	140°
(4) Hot	1.5	150°	30°

$\Delta T_{min} = 10°C$

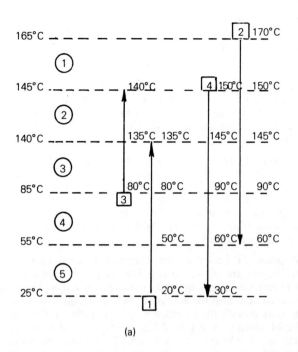

(a)

	INTERVAL No. i	$T_i - T_{i+1}$ (°C)	$\Sigma CP_{cold} - \Sigma CP_{hot}$ (kW/°C)	ΔH_i (kW)	SURPLUS OR DEFICIT
T_1 = 165°C					
	1	20	−3.0	−60	SURPLUS
T_2 = 145°C					
	2	5	−0.5	−2.5	SURPLUS
T_3 = 140°C					
	3	55	+1.5	+82.5	DEFICIT
T_4 = 85°C					
	4	30	−2.5	−75	SURPLUS
T_5 = 55°C					
	5	30	+0.5	+15	DEFICIT
T_6 = 25°C					

(b)

Figure 2.14—Temperature interval analysis

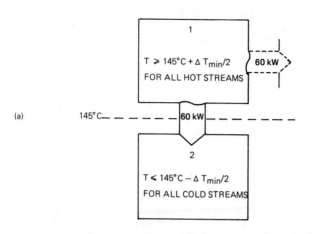

(a)

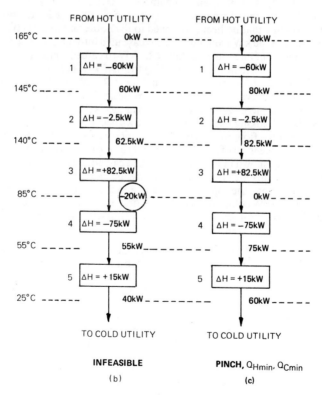

Figure 2.15—The heat cascade principle—target prediction by "problem table" analysis

27

135°C to 140°C. Setting up the intervals in this way *guarantees* that full heat interchange within any interval is possible. Hence, each interval will have either a net surplus or net deficit of heat as dictated by enthalpy balance, *but never both*. This is shown in Figure 2.14 (b). Knowing the stream population in each interval (from Figure 2.14 (a)), enthalpy balances can easily be calculated for each according to: —

$$\Delta H_i = (T_i - T_{i+1})(\Sigma CP_C - \Sigma CP_H)_i$$

for any interval *i*. The last column in Figure 2.14 (b) indicates whether an interval is in heat surplus or heat deficit. It would therefore be possible to produce a feasible network design based on the assumption that all "surplus" intervals rejected heat to cold utility, and all "deficit" intervals took heat from hot utility. However, this would not be very sensible, because it would involve rejecting and accepting heat at inappropriate temperatures.

We now, however, exploit a key feature of the temperature intervals. Namely, *any heat available in interval i is hot enough to supply any duty in interval i + 1*. This is shown in Figure 2.15 (a), where intervals 1 and 2 are used as an illustration. Instead of sending the 60 kW of surplus heat from interval 1 into cold utility, it can be sent down into interval 2. It is therefore possible to set up a heat "cascade" as shown in Figure 2.15 (b). Assuming that no heat is supplied to the hottest interval (1) from hot utility, then the surplus of 60 kW from interval 1 is cascaded into interval 2. There it joins the 2.5 kW surplus from interval 2, making 62.5 kW to cascade into interval 3. Interval 3 has a 82.5 kW deficit, hence after accepting the 62.5 kW it can be regarded as passing on a 20 kW deficit to interval 4. Interval 4 has a 75 kW surplus and so passes on a 55 kW surplus to interval 5. Finally, the 15 kW deficit in interval 5 means that 40 kW is the final cascaded energy to cold utility. This in fact is the net enthalpy balance on the whole problem. Looking back at the heat flows between intervals in Figure 2.15 (b), clearly the negative flow of 20 kW between intervals 3 and 4 is thermodynamically infeasible. To make it just feasible (*i.e.* equal to zero), 20 kW of heat must be added from hot utility as shown in Figure 2.15 (c), and cascaded right through the system. By enthalpy balance this means that all flows are increased by 20 kW. The net result of this operation is that the minimum utilities requirements have been predicted, *i.e.* 20 kW hot and 60 kW cold. Further, the position of the pinch has been located. This is at the 85°C interval boundary temperature (*i.e.* hot streams at 90°C and cold at 80°C) *where the heat flow is zero*.

Compare the results obtained by this approach to the results from the composite curves, as shown in Figure 2.16. The same information is obtained, but the Problem Table provides a simple framework for numerical analysis. For simple problems it can be quickly evaluated by hand. For larger problems, it is easily implemented on the computer. It can also be adapted for the case where the value of ΔT_{min} allowed depends on the streams matched, and is not simply a "global" value (see under section 2.3). Finally it can be adapted to cover other cases where simplifying assumptions (for example, CP = constant) are invalid, as will be shown later.

With the Problem Table algorithm, the engineer has a powerful targeting technique at his or her fingertips. Data can be quickly extracted from flowsheets and analysed to see whether the process is nearing optimal, or whether significant scope for energy saving exists. This provides enormous stimulus to break away from the "learning

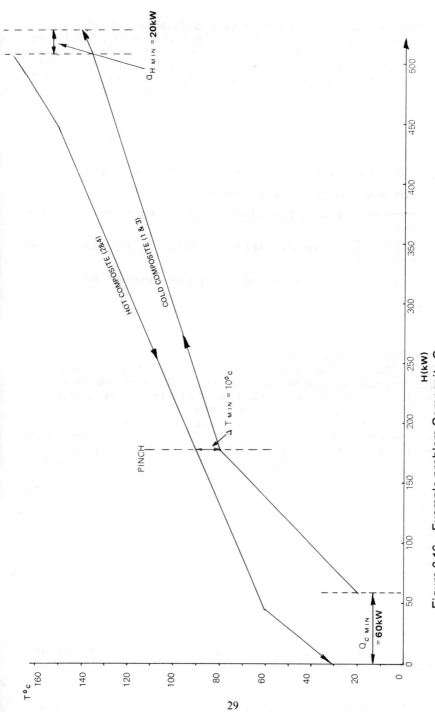

Figure 2.16—Example problem Composite Curves

29

curve", as illustrated in Figure 2.1. It can also be used to settle quickly disputes along the lines of "to integrate, or not to integrate?" Processes often fall into distinct sections ("A" and "B",) by reason of layout or operability considerations. The question is often "can significant savings be made by cross integration?" The Problem Table algorithm can be applied to areas A and B separately, and then to all the streams in A and B together. The results of the analysis will quickly settle the question. For example, if the answer is: −

A	alone	: 10% savings in total fuel bill possible
B	alone	: 5% savings in total fuel bill possible
A & B together		: 30% savings possible

then there is a 15% energy incentive for cross-integrating areas A and B.

To summarise this section on energy targeting:

● Composite curves give conceptual understanding of how energy targets can be obtained.

● The Problem Table gives the same results (including the "pinch" location) more easily.

● Energy targeting is a powerful design and "process integration" aid.

2.2.3. Simple design for maximum energy recovery

2.2.3.1. The significance of the "pinch"—Figure 2.17 (a) shows the composite curves for a multi-stream problem dissected at the pinch. "Above" the pinch (*i.e.* in the region to the right) the composite hot transfers all its heat into the composite cold leaving utility *heating only* required. The region above the pinch is therefore a *heat sink*, with heat flowing into it but no heat flowing out. Conversely below the pinch *cooling only* is required and the region is therefore a *heat source*. The problem therefore falls into two thermodynamically distinct regions, as indicated by the enthalpy balance envelopes in Figure 2.17 (b). Heat $Q_{H(MIN)}$ flows into the problem above the pinch and $Q_{C(MIN)}$ out of the problem below, but *the heat flow across the pinch is zero*. This result was observed in the description of the Problem Table algorithm in the previous sub-section. It follows that any network design that transfers heat α across the pinch must, by overall enthalpy balance, require α more than minimum from hot and cold utilities, as shown in Figure 2.17 (c). As a corollary, any utility cooling α above the pinch must incur extra hot utility α, and *vice versa* below the pinch. So for the designer wishing to produce a minimum utility design, the firm message is: −

● Don't transfer heat across the Pinch.

● Don't use cold utilities above.

● Don't use hot utilities below.

The decomposition of the problem at the pinch turns out to be very useful when it comes to network design (Linnhoff and Hindmarsh, 1982).

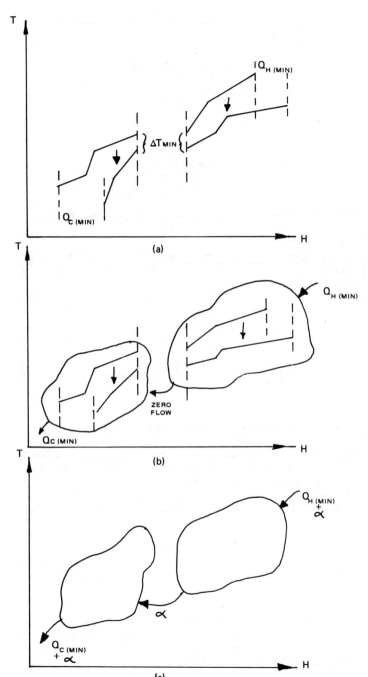

Figure 2.17—The Pinch decomposition

31

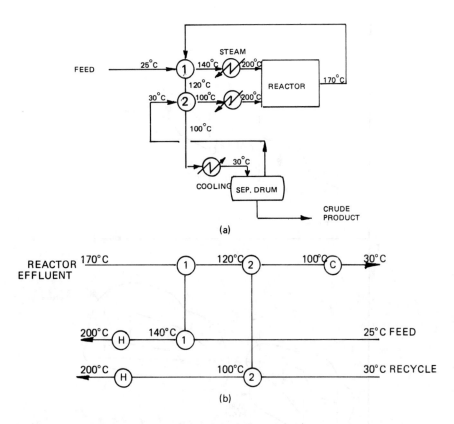

Figure 2.18—Heat exchanger network representation

2.2.3.2 Network representation—The heat exchanger network from the flowsheet in Figure 2.18 (a) can be represented in the "grid" form introduced by Linnhoff and Flower (1978); see Figure 2.18 (b). The advantage of this representation is that the heat exchange matches 1 and 2 (each represented by two circles joined by a vertical line in the grid) can be placed in either order without redrawing the stream system. In the flowsheet representation, if it were desired to match recycle against the hottest part of the reactor effluent, the stream layout would have to be redrawn. Also, the grid represents the countercurrent nature of the heat exchange, making it easier to check exchanger temperature feasibility. Finally, the pinch is easily represented in the grid (as will be shown in the next sub-section), whereas it cannot be represented on the flowsheet.

2.2.3.3. Design for best energy recovery—The data in Table 2.1 were analysed by the Problem Table method in sub-section 2.2.2.3 with the result that the minimum utility requirements are 20 kW hot and 60 kW cold. The pinch occurs where the hot streams are at 90°C and the cold at 80°C.

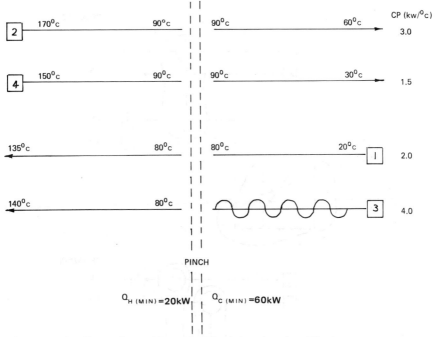

					CP (kw/^{0}c)
2	170^{0}c ——————— 90^{0}c		90^{0}c ——————————— 60^{0}c		3.0
4	150^{0}c ——————— 90^{0}c		90^{0}c ——————————— 30^{0}c		1.5
	135^{0}c ——————— 80^{0}c		80^{0}c ——————————— 20^{0}c	1	2.0
	140^{0}c ——————— 80^{0}c		∿∿∿∿∿	3	4.0

PINCH

$Q_{H\,(MIN)}=20kW$ $Q_{C\,(MIN)}=60kW$

Figure 2.19—Example problem stream data, showing Pinch

The grid structure for the problem is shown in Figure 2.19, with the pinch represented as a vertical dotted line. Above the pinch the hot streams are cooled from their supply temperatures to their pinch temperature, and the cold streams heated from their pinch temperature to their target temperatures. Below the pinch the position is reversed with hot streams being cooled from the pinch to target temperatures and cold streams being heated from supply to pinch temperature. Notice that stream number 3 starts at the pinch. In fact in problems where the streams all have constant *CP*s, the pinch is always caused by the entry of a stream, either hot or cold.

We remember now that above the pinch, if best performance is to be obtained, no utility cooling should be used. This means that above the pinch, *all hot streams must be brought to pinch temperature by interchange against cold streams*. We must therefore start the design at the pinch, finding matches that fulfil this condition. In this example, above the pinch there are two hot streams at pinch temperature, therefore requiring two "pinch matches". In Figure 2.20 (a) a match between streams 2 and 1 is shown, with a T/H plot of the match shown in inset. (Note that the stream directions have been reversed so as to mirror the directions in the grid representation.) Because the *CP* of stream 2 is greater than that of stream 1, as soon as any load is placed on the match, the ΔT in the exchanger becomes less than ΔT_{min} at its hot end. The exchanger is clearly infeasible and therefore we must look for another match. In Figure 2.20 (b), streams 2 and 3 are matched, and now the relative gradients of the T/H plots mean that putting load on the exchanger opens up the ΔT.

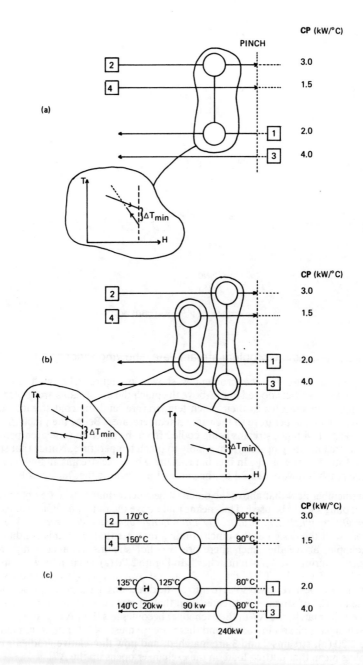

Figure 2.20—Example problem hot end design

34

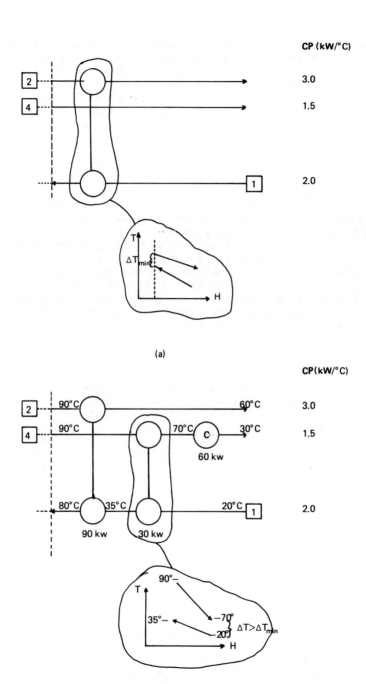

(a)

(b)
Figure 2.21—Example problem cold end design

This match is therefore acceptable. If it is put in as a firm design decision, then stream 4 must be brought to pinch temperature by matching against stream 1 (*i.e.* this is the only option remaining for stream 4). Looking at the relative sizes of the CPs for streams 4 and 1, the match is feasible ($CP_4 < CP_1$). There are no more streams requiring cooling to pinch temperature and so we have found a feasible pinch design. It is the only feasible pinch design because only two pinch matches are required. Summarising, in design immediately above the pinch, it is required to meet the criterion:

$$CP_{\text{HOT}} \leqslant CP_{\text{COLD}}$$

Having found a feasible pinch design it is necessary to decide on the match heat loads. The recommendation is "maximise the heat load so as to completely satisfy one of the streams". This ensures minimum number of units employed as will be shown in the next sub-section. So, since stream 2 above the pinch requires 240 kW of cooling and stream 3 above the pinch requires 240 kW of heating, co-incidentally the 2/3 match is capable of satisfying both streams. However, the 4/1 match can only satisfy stream 4, having a load of 90 kW and therefore heating up stream 1 only as far as 125°C. Since both hot streams have now been completely exhausted by these two design steps, stream 1 must be heated from 125°C to its target temperature of 135°C by external hot utility as shown in Figure 2.20 (c). This amounts to 20 kW, as predicted by the Problem Table analysis. This is no coincidence! The design has been put together obeying the constraint of not transferring heat across the pinch (the "above the pinch" section has been designed completely independently of the "below the pinch" section) and not using utility cooling above the pinch.

Below the pinch, the design steps follow the same philosophy, only with design criteria that mirror those for the "above the pinch" design. In Figure 2.21 (a) the stream system below the pinch is shown. Now, it is required to bring cold streams to pinch temperature by interchange with hot streams, since we do not want to use utility heating below the pinch. In this example, only one cold stream exists below the pinch which must be matched against one of the two available hot streams. The match between streams 1 and 2 is feasible, as shown by the inset diagram in Figure 2.21 (a) because the CP of the hot stream is greater than that of the cold. The other possible match (stream 1 with stream 4) is not feasible. Immediately below the pinch, the necessary criterion is:

$$CP_{\text{HOT}} \geqslant CP_{\text{COLD}}$$

... which is the inverse of the criterion for design immediately above the pinch.

Maximising the load on this match satisfies stream 2, the load being 90 kW. The heating required by stream 1 is 120 kW and therefore 30 kW of residual heating, to take stream 1 from its supply temperature of 20°C to 35°C, is required. Again this must come from interchange with a hot stream, the only one now available being stream 4. Although the CP inequality does not hold for this match, the match is feasible because *it is away from the pinch*. That is to say, it is not a match that has to bring the cold stream up to pinch temperature. So, as shown in inset in Figure 2.21 (b), the match does not become infeasible. Putting a load of 30 kW on this match leaves residual cooling of 60 kW on stream 4 which must be taken up by cold utility. Again, this is as predicted by the Problem Table analysis.

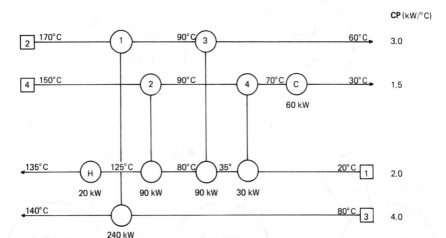

CP (kW/°C)

Figure 2.22—Example problem completed design

Putting the "hot end" and "cold end" designs together gives the completed design shown in Figure 2.22. It achieves best possible energy performance for a ΔT_{min} of 10°C incorporating four exchangers, one heater and one cooler. In other words, six units of heat transfer equipment in all.

Summarising, this design was produced by:

● Dividing the problem at the pinch, and designing each part separately.

● Starting the design *at* the pinch and moving away.

● Immediately adjacent to the pinch, obeying the constraints:

$$CP_{HOT} \leqslant CP_{COLD} \text{ (above)}$$

$$CP_{HOT} \geqslant CP_{COLD} \text{ (below)}$$

● Maximising exchanger loads.

● Supplying external heating only above the pinch, and external cooling only below the pinch.

These are the basic elements of the "pinch design method" of Linnhoff and Hindmarsh (1982). They will be further elaborated in later sub-sections.

2.2.3.4. A word about design strategy—The method just described does not follow the traditional intuitive method for heat exchanger network design. Left to his own devices, the engineer normally starts to design from the hot end, working his way towards the cold. However, the "pinch design method" *starts the design where the problem is most constrained*. That is, at the pinch. The thermodynamic constraint of the pinch is "used" by the designer to help him identify matches that must be made in order to produce efficient designs. Where it is possible to identify options at the pinch

37

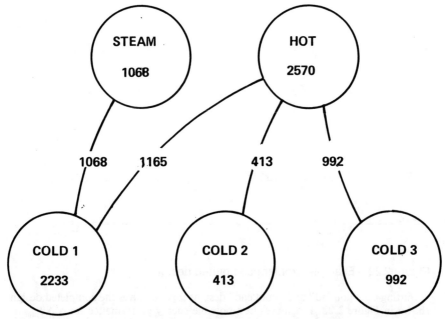

Figure 2.23—Illustration of minimum number of units design

(and this will be discussed later), the designer may choose the one he likes for control, layout, safety, or other reasons, and still be sure that an energy-efficient design will result.

2.2.4. Trading off energy against capital

2.2.4.1. *What is the "minimum number of units?"*—The capital cost of chemical processes tends to be dominated by the number of items on the flowsheet. This is certainly true of heat exchanger networks and there is a strong incentive to reduce the number of matches between hot and cold streams.

Referring back to the flowsheet in Figure 2.5, three exchangers, two heaters, and one cooler are used in the design, making six units in all. Is this the minimum number, or could the designer have managed with fewer units?

As previously described, the bare flow diagram in Figure 2.6 shows that there are four separate process streams to consider. The target energy performance for this system as calculated by the Problem Table method shows that only heating is required, and no cooling. Straight away then, we know that the cooler is surplus to requirement! Figure 2.23 shows the heat loads on the one hot stream and three cold streams written within circles representing the streams. The predicted hot utility load is shown similarly. Note that the total system is in enthalpy balance, *i.e.* the total hot plus utility is equal to the total cold. If we assume that temperature constraints will allow any match to be made, then we can start putting in matches as follows. Matching STEAM with COLD 1 and maximising the load completely satisfies or "ticks off" STEAM, leaving 1165 units of heating required by COLD 1. Matching COLD 1 with HOT and maximising the load on this match so that it ticks off the 1165

38

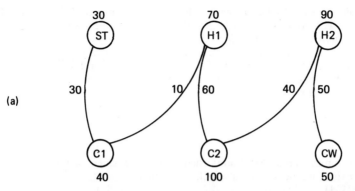

(a)

NUMBER OF UNITS IS ONE LESS THAN THE NUMBER OF STREAMS
INC. UTILITIES

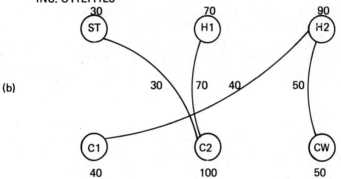

(b)

SAME PRINCIPLE FOR SEPARATE COMPONENTS –'SUBSET EQUALITY'

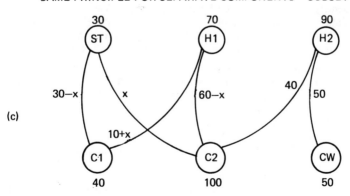

(c)

ONE UNIT MORE FOR EVERY LOOP

Figure 2.24—Principles of subsets and loops

39

residual requirement on COLD 1, leaves 1405 residual heat available from HOT. Matching HOT with both COLD 2 and COLD 3 and maximising loads then ticks off both the residual on HOT and the requirements on COLD 2 and COLD 3. So following the principle of maximising loads, *i.e.* "ticking off" stream or utility loads or residuals, leads to a design with a total of four matches. This is in fact the minimum for this problem. Notice that it is one less than the total number of streams plus utilities in the problem.

Thus:

$$u_{min} = N - 1$$

where u_{min} = minimum number of units (including heaters and coolers)
and N = total number of streams (including utilities)
(Hohmann, 1971).

It is possible to produce a design for this sytem with four units, as shown in the Introduction of this Guide, Figure 1.1. In fact, it is normally possible in heat exchanger network design to find a u_{min} solution, as will be shown.

Certain refinements to this formula are required, however. In Figure 2.24 (a), a problem having two hot streams and two cold streams is shown. In this case, both utility heating and utility cooling are required. Putting in matches as before by ticking off loads or residuals leads to a design with $N - 1$ units. However, in Figure 2.24 (b) a design is shown having one unit less. The reason why the number appears less than minimum is not hard to see. Whilst overall the problem is in enthalpy balance, the subset of streams H2, C1, and CW is itself in enthalpy balance. Similarly ST, H1, and C2 are in enthalpy balance (which they must be if the total problem is in balance). What this means is that for the given data set we can design two completely separate networks, with the formula $u_{min} = N - 1$ applying to each individually. The total for the overall system is therefore $(3-1) + (3-1) = 4$ units, or one less than in Figure 2.24 (a). This situation is termed "subset equality", *i.e.* for the given data set it is possible to identify two subsets which by enthalpy balance can form separate networks. The data set is said to comprise two "components". Since the flowsheet designer is in control of the size of the heat loads in his plant, it is sometimes possible to deliberately change loads so as to force subset equality and thus save a unit.

Finally, in Figure 2.24 (c) a design is shown having one unit more than the design in Figure 2.24 (a), the new unit being the match between ST and C2. The extra unit introduces what is known as a "loop" into the system. That is, it is possible to trace a closed path through the network. Starting, say, at the hot utility ST, the loop can be traced through the connection to C1, from C1 to H1, from H1 to C2, and from C2 back to ST. The existence of the loop introduces an element of flexibility into the design. Suppose the new match, which is between ST and C2, is given a load of X units. Then by enthalpy balance, the load on the match between ST and C1 has to be $30 - X$, between C1 and H1 $10 + X$, and between H1 and C2 $60 - X$. Clearly X can be anything up to a value of 30, when the match between ST and C1 disappears. The flexibility in design introduced by loops is sometimes useful, particularly in "revamp" studies.

The features discussed with Figure 2.24 are described by a theorem from graph

40

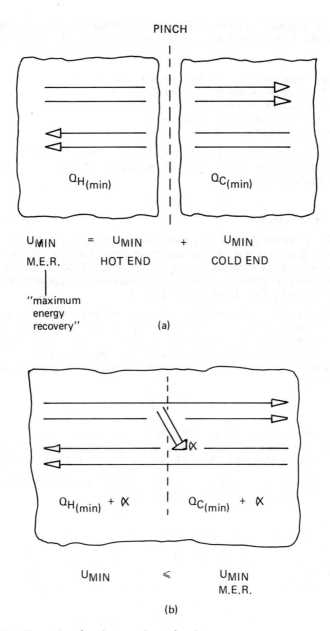

Figure 2.25—Targeting for the number of units

theory in mathematics, known as Euler's General Network Theorem. This theorem, "translated" into the terminology of heat exchanger networks, states that

$$u = N + L - s$$

where u = number of units (including heaters and coolers),
N = number of streams (including utilities)
L = number of loops
s = number of separate components.
(Linnhoff, et al., 1979).

Normally we want to avoid extra units, and so design for $L = 0$. Also, unless we are lucky, there will be no subset equality in the data set and hence $s = 1$. This then leads to the targeting equation

$$u_{min} = N - 1$$

introduced previously.

2.2.4.2. *Targeting for the minimum number*—Figure 2.25 (a) shows how the targeting equation is applied to a "maximum energy recovery" (MER) design. The pinch divides the problem into two thermodynamically independent regions. Since the regions are independent, the targeting formula must be applied to each separately as shown (Linnhoff and Hindmarsh, 1982).

The total for the whole problem, "$u_{min, MER}$", is then the sum of the u_{mins} for each region. Suppose, however, that α units of heat are transferred across the pinch as shown in Figure 2.25 (b), thus increasing the hot and cold utilities by α. Now, the regions are no longer thermodynamically independent. Applying the targeting formula to the "energy slack" total problem, *i.e.* ignoring the pinch, leads to the conclusion that

$$u_{min} \leqslant u_{min\ MER}$$

This is because in targeting for the MER design, streams that cross the pinch are counted twice. The conclusion is that there is a trade-off between energy recovery and number of units employed. This trade-off will be more fully described later.

Referring back to the example problem shown in Figure 2.19 and applying the targeting formula to the hot and cold ends, we obtain:

$$u_{min\ MER} = (5 - 1) + (4 - 1) = 7$$

However, the final design shown in Figure 2.22 has only six units. The reason is the co-incidence of data mentioned in the decription of the hot-end design. Above the pinch as shown in Figure 2.20 (c), streams 2 and 3 form a subset, allowing the hot end to be designed with three units rather than four ($s = 2$ in Euler's equation). Applying the targeting formula to the whole problem ignoring the pinch gives:

$$u_{min} = (6 - 1) = 5$$
$$(4 \text{ streams } + 2 \text{ utilities})$$

Hence by transferring energy across the pinch, the scope for reducing the number of units is 1.

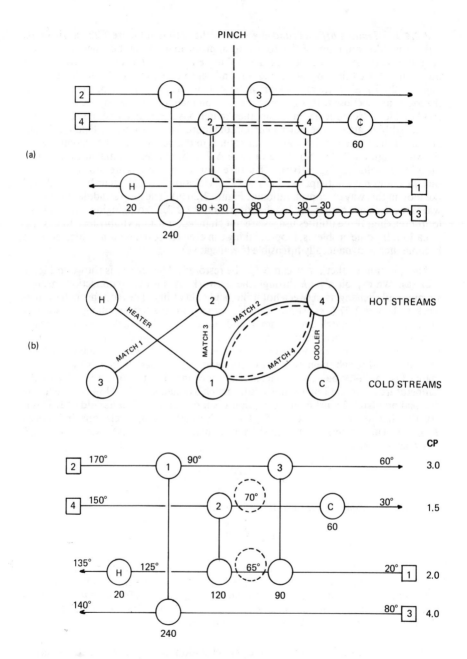

Figure 2.26—Example problem loop-breaking

43

2.2.4.3. Trading off units and energy—As the design in Figure 2.22 has six units, rather than the minimum of five for the total problem ignoring the pinch, there must be a loop in the system (as discussed with Figure 2.24 (c)). The loop is shown traced out with a dotted line in Figure 2.26 (a) and reproduced in the alternative form in Figure 2.26 (b). Since there is a loop in the system, the load on one of the matches in the loop can be chosen. If we choose the load on match 4 to be zero, *i.e.* we subtract 30 kW of load from the design value, then match 4 is eliminated and the 30 kW must be carried by match 2, the other match in the loop. This is shown in Figure 2.26 (a). Having shifted loads in this way, temperatures in the network can be recomputed as shown in Figure 2.26 (c). Now, the value of ΔT at the cold end of match 2 is less than the allowed value ($\Delta T_{min} = 10°C$). The offending temperatures are shown circled. In fact we could have anticipated that a "ΔT_{min} violation" would occur by "breaking" the loop in this way by consideration of Figure 2.26 (a). The loop straddles the pinch, where the design is constrained as described under 2.2.3.3. So changing this design by loop-breaking, if the utilities usages are not changed, must inevitably lead to a ΔT_{min} violation. In some problems, loop-breaking can even cause temperature differences to become thermodynamically infeasible (*i.e.* negative).

The question is, then, how can ΔT_{min} be restored? The answer is shown in Figure 2.27 (a). We exploit a *path* through the network. A path is a connection through streams and exchangers between hot utility and cold utility. The path through the network in Figure 2.27 (a) is shown dotted, going from the heater, along stream 1 to match 2, through match 2 to stream 4, and along stream 4 to the cooler. If we add a heat load X to the heater, then by enthalpy balance the load on match 2 must be reduced by X and the load on the cooler increased by X. Effectively we have "pushed" extra heat X through the network, thereby reducing the load on match 2 by X. Now match 3 is not in the path, and so its load is not changed by this operation. Hence the temperature of stream 1 on the hot side of match 3 remains at 65°. However, reducing the load on match 2 must increase T_2, thus opening out the ΔT at its cold end. This is exactly what we need to restore ΔT_{min}! There is clearly a simple relationship between T_2 and X. The temperature fall on stream 4 in match 2 is $(120 - X)$ divided by the CP of stream 4. Hence,

$$150° - \frac{(120 - X)}{1.5} = T_2$$

Alternatively, applying the same logic to the cooler,

$$30° + \frac{(60 + X)}{1.5} = T_2$$

Since $\Delta T_{min} = 10°$ we want to restore T_2 to 75°. Solving either of the above equations with $T_2 = 75°$ yields $X = 7.5$ kW. Since ΔT_{min} is exactly restored, 7.5 kW must be the minimum energy sacrifice required to produce a u_{min} solution from the u_{min} MER solution. The "relaxed" solution is shown in Figure 2.27 (b), with the temperature between the heater and match 2 on stream 1 computed.

In summary on this subject of "energy relaxation", the procedure for reducing units at minimum energy sacrifice is:

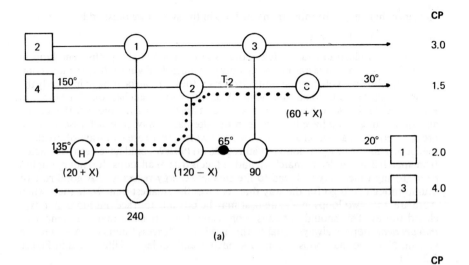

(a)

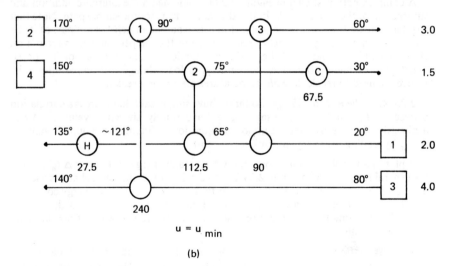

$$u = u_{min}$$

(b)

Figure 2.27—Example problem energy relaxation

- Identify a loop (across the pinch).
- Break it by subtracting and adding loads.
- Recalculate network temperatures and identify the ΔT_{min} violations.
- Find a relaxation path and formulate $T = f(X)$.
- Restore ΔT_{min}

and since there might be more than one loop in the system we must add

● Repeat for other loops.

Figure 2.28 illustrates just a few more aspects of loops and paths. The loop in Figure 2.26 (a) is a simple one involving only two units. In Figure 2.28 (a) a more complex one is shown involving four units. However, the loop can be broken in exactly the same way, that is adding and subtracting X on alternative matches round the loop. In Figure 2.28 (a) the loop breaks when X equals either L_1 or L_4. Note that the adding and subtracting could have been done in the alternative way, in which case it would break when X equals either L_2 or L_3. In other words, there are two ways of breaking the loop. This is true of the loop in Figure 2.26 (a) (90 kW could have been subtracted from match 2 and added to match 4), and in fact is true of all loops. It is not possible *a priori* to say which way will lead to the smallest energy relaxation. A good rule of thumb though is to go for the way that removes the smallest unit. Note that, when there are, say, two loops in a system, it may be possible to trace out more than two closed routes. This should not cause confusion if it is realised that the number of *independent* loops is always equal to the number of "excess" units ($> N-1$) in the system. Note too that loops can include heaters and coolers, as illustrated in Figure 2.28 (b).

A complex path is shown in Figure 2.28 (c), and again the alternate addition and subtraction of the load X works in just the same way as for the simple path. Note that although the path goes through match 1 in this example, match 1 is not part of it. Its load is not changed by the energy relaxation, but the temperatures on stream 4 on either side of it *are* changed. When a similar situation occurs within a loop it is possible for the exchanger that does not undergo a load change to become infeasible. Hence the need to recalculate *all* temperatures after loop-breaking.

2.2.4.4. The Role of ΔT_{min}—So far we have simply said that there is a correlation between ΔT_{min} and utilities usages, and have simply chosen a value of ΔT_{min} arbitrarily. Now we examine the important question of finding the economic value of ΔT_{min} by trading off energy against capital cost.

A plot of annualised capital cost *versus* ΔT_{min}, and energy cost *versus* ΔT_{min}, takes the general form shown in Figure 2.29 (a). As driving forces increase, characterised by increasing ΔT_{min}, capital cost falls (from infinity at $\Delta T_{min} = 0$) and energy cost rises. The total cost therefore passes through a minimum. There is no straightforward method of locating this minimum precisely, but there are two ways of obtaining an approximate value.

The first method depends on an observation made by Hohmann (1971) that all networks featuring maximum energy recovery show similar surface area requirements, and that this area is approximately equal to the "total minimum area" calculated from the composite curves. This area can be estimated assuming a "mean" heat transfer coefficient for the whole problem. By dividing up the composite curves into simple countercurrent sections, and then applying the usual heat transfer equation $Q = U.A.\Delta T_{LM}$ to each, a total area for each section is obtained. The sum of the areas calculated in this way is the total minimum area. By weighting this area according to the number of units required (from the units targeting procedure), an approximate total network cost can be produced. This then allows capital cost to be traded against

46

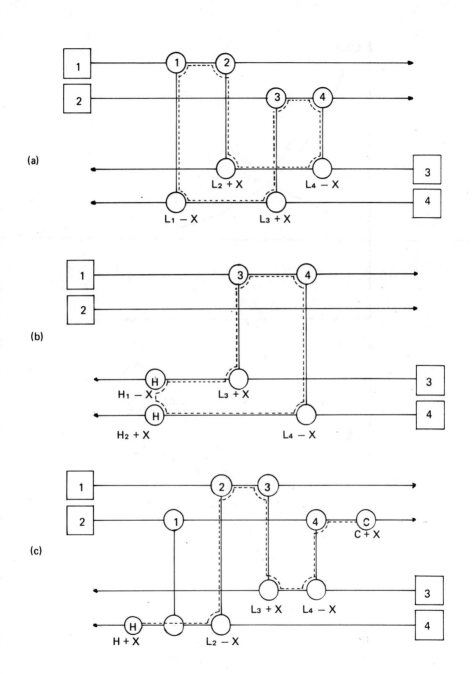

Figure 2.28—Complex loops and paths

47

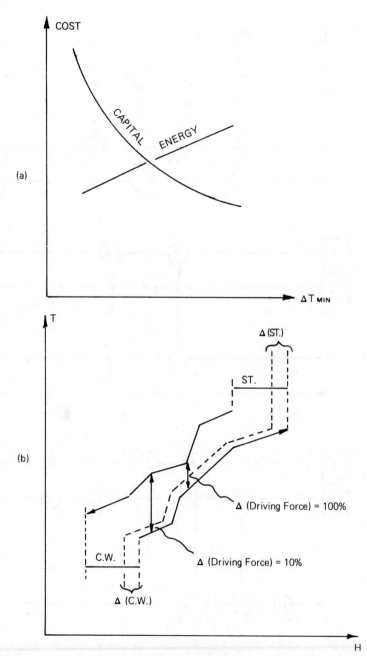

Figure 2.29—Assessment of the economic ΔT_{min}

utilities cost. The drawback of this method is that it does not allow for widely differing U-values.

An alternative method is illustrated in Figure 2.29 (b). Where problems have a "severe" pinch, changes in ΔT_{min} have a much more marked effect on capital cost at the pinch than elsewhere in the problem. Thus if ΔT_{min} is doubled from 10°C to 20°C, then the change in driving force at the pinch is 100%. However, away from the pinch where driving forces are, say, of the order of 100°C, an increase in ΔT_{min} of 10°C means only a 10% increase in driving force. This suggests a quick method for finding the optimum ΔT_{min}. By just considering the design in the region of the pinch (obtained by the Pinch Design Method) and assuming that changes in ΔT_{min} affect the

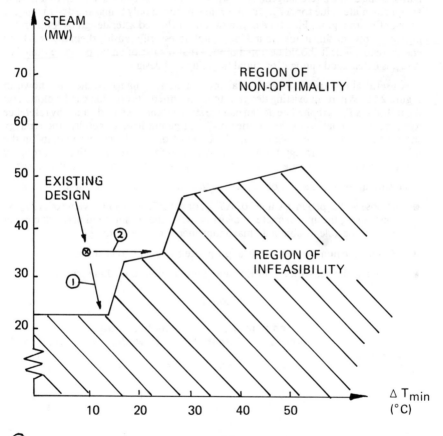

① Decrease Steam Usage (& Reduce Capital Cost?)
② Reduce Capital Cost.

Figure 2.30—Strategy for design changes

49

cost of the Pinch Design only, capital cost can be traded against utilities cost without considering the complete network.

Having seen how the value of ΔT_{min} is obtained, the reader will appreciate that when it comes to design, ΔT_{min} is not "gospel". Referring back to Figure 2.27 (a), after loop breaking, energy was relaxed to restore ΔT_{min} in match 2. However, the design was thermodynamically feasible prior to relaxation (ΔT at cold end of match 2 = 5°C). The correct degree of relaxation is determined by capital energy trade-off in a manner similar to that shown in Figure 2.29 (a). That is, as energy is relaxed along the path, driving forces open up and so overall, capital cost of the affected exchangers falls. (Note, though, that the cost of the utilities exchangers increases, and in complex paths, some of the interchangers' costs will increase). Hence it can be seen that finding an approximate value for ΔT_{min} for the whole problem only brings the design into the near-optimum region. Fine tune optimisation is achieved after design by loop breaking (or simply shifting loads around loops, not necessarily causing them to break) and energy relaxation. It should be noted though that the use of an "experience value" of ΔT_{min} is often good enough for initial targeting and design.

A useful alternative strategy is available for design change studies, as shown in Figure 2.30. Where an existing design is to be modified, its position can be plotted on the utilities ΔT_{min} graph. The design change strategy can then be decided by reference to its position relative to the utilities ΔT_{min} targeting line. Generally, the strategy should be to evolve the design in the direction of one of the concave points on the utilities ΔT_{min} line. Strategy 1 in Figure 2.30 is therefore to reduce utilities at roughly constant capital cost. Strategy 2 is to reduce captial cost at constant utilities usage.

To summarise this section on capital *versus* energy tradeoffs, the designer should:

● Choose an "experience value" of ΔT_{min}, or get an approximate value of ΔT_{min} by considering the approximate tradeoff between energy and total network cost, in order to be able to obtain a "maximum energy recovery" design.

● Reduce the number of units as appropriate by "energy relaxation".

● Optimise the network by shifting load around loops and along paths.

Table 2.2

Process stream no. type	Temperature		Heat capacity flowrate, C_p $10^4/(Btu)/(h)(°F)$	Heat load, Q 10^4 Btu/h
	Supply T_S °F	Target T_T °F		
1 *cold*	200	400	1.6	320.0
2 *cold*	100	430	1.6	528.0
3 *hot*	590	400	2.376	451.4
4 *cold*	300	400	4.128	412.8
5 *hot*	471	200	1.577	427.4
6 *cold*	150	280	2.624	341.1
7 *hot*	533	150	1.32	505.6

$\Delta T_{min} = 20°F$ Data are for problem 7SP2

An interesting footnote can be added. The capital/energy cost graph in Figure 2.29 (a) shows the capital cost line decreasing monotonically with increasing energy. However, if energy is increased far enough, *the capital cost can start to increase again* up to a maximum value (when all duties are performed by utilities). This is because the effect of increasing heat load starts to dominate over the effect of increasing driving forces. Past experience with the techniques described in this Guide shows that many "state-of-the-art" processes have designs in the rising region of the capital cost graph. Hence, *application of the techniques to state-of-the-art processes can lead to both energy and capital savings*. This is exemplified by a simple "literature" problem "7SP2" first introduced by Masso and Rudd (1969), the data for which are given in Table 2.2. Figure 2.31 (a) shows the non-integrated solution, having operating costs of $250 838 per year (according to the criteria given by Masso and Rudd) and annualised capital cost of $4937 per year. The optimally integrated solution in Figure 2.31 (b) has operating and capital costs of $24 077 per year and $4180 per year respec-

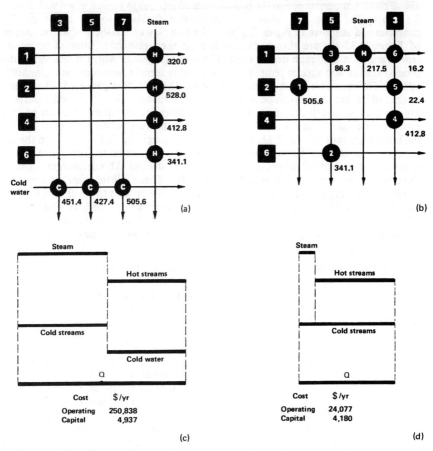

Figure 2.31—Comparison of integrated and non-integrated designs

51

tively. Figures 2.31 (c) and (d) illustrate the big reduction in total network heat duty brought about by integration, which more than compensates for the reduction in driving forces.

2.2.5. *Multiple utilities*

So far in the discussion, the possibility of using many utility levels has not been considered. This is clearly unrealistic, and a method is required to deal with the problem. In the Problem Table algorithm for energy targeting, the implicit assumption is that hot utility is hot enough to perform any required heating duty, and that cold utility is cold enough for any cooling duty. The Problem Table can however be adapted to cater for the multiple utilities case.

2.2.5.1. Understanding process sources and sinks—the "Grand Composite Curve"—Figure 2.32 (a) shows a schematic heat cascade diagram (of the type introduced in Figure 2.15 (c)). The pinch divides the cascade into the process heat sink above pinch temperature, and the process heat source below (as discussed with Figure 2.17 (b)). The heat sink above the pinch and the heat source below can be characterised as shown in Figure 2.32 (b). The heat flows from the cascade on the left of Figure 2.32 (b) are shown plotted against their respective interval boundary temperatures in the graph on the right. The result is a graph which characterises the process source and sink in temperature—enthalpy terms (Townsend and Linnhoff, 1982a). The graph above the pinch represents a sort of "net process cold stream" against which hot utilities can be matched in countercurrent flow. See the top parts of Figures 2.33 (a) and (b). Similarly the graph below the pinch represents a sort of "net process hot stream" against which cold utilities can be matched countercurrently. See the bottom parts of Figures 2.33 (a) and (b). Note though that below the pinch the net process hot stream (or "process source profile" as we will call it from now on) runs in the reverse direction to that previously used for hot streams. This arises simply out of the construction method shown in Figure 2.32 (b). There is little point, however, in dissecting the graph to turn the process source profile around to make it consistent with the more usual hot stream representation. In fact there is a positive advantage in leaving it as it is. The point of gradient change at zero heat flow where the process source profile and the process sink profile (*i.e.* the net process cold stream) meet, clearly represents the pinch.

In the shaded regions shown in Figure 2.33 (a) the analogy with single stream heat exchange breaks down. Above the pinch, section AB of the graph represents the heat surplus temperature interval No. 2 (Figure 2.32 (b)). It therefore represents a "local" heat source in the midst of the net process heat sink. Similarly below the pinch, section FE represents a local heat sink (interval No. 6) in the midst of the net process heat source. By the heat cascading principle, the heat available in AB is hot enough to be transferred into the process sink anywhere between the pinch and point B. However, if it is transferred into CB, then it is transferred at the minimum possible driving force. This allows the best possible use of low temperature hot utilities (*i.e.* over the region between the pinch and point C). Similarly, if below the pinch the heat required by FE is taken from DE, the best possible use can be made of high temperature cold utilities. Thus in the shaded regions in Figure 2.33 (a), the process effectively takes care of itself, *i.e.* it is in enthalpy balance. It is only those parts of the graph outside these regions that represent the process source and sink profiles.

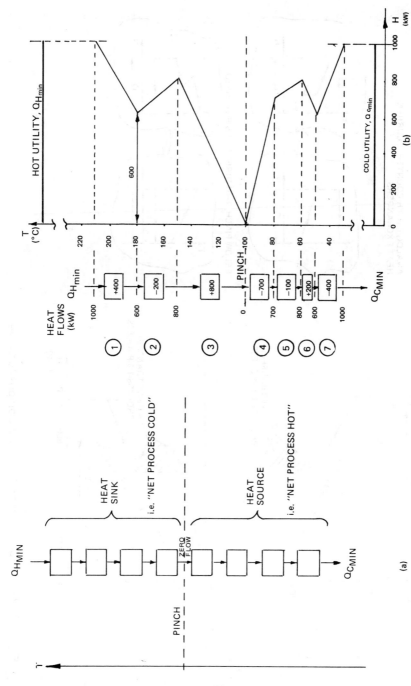

Figure 2.32—Construction of the "Grand Composite Curve"

53

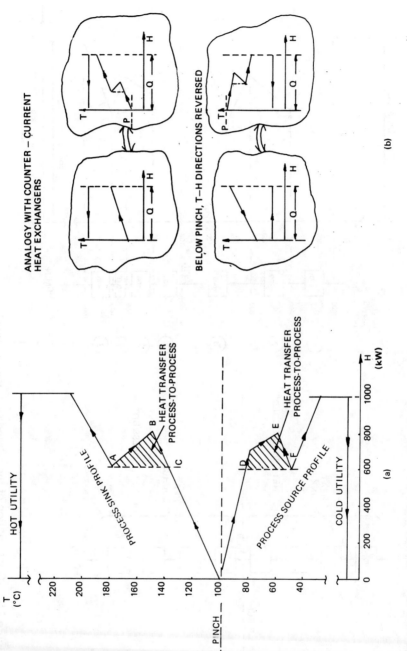

Figure 2.33—Interpretation of the Grand Composite Curve.

54

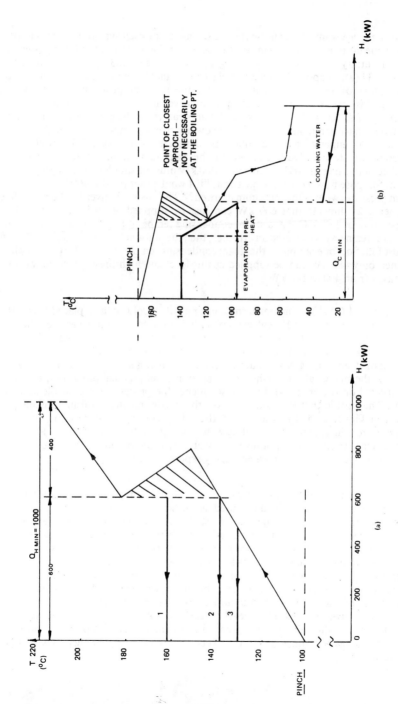

Figure 2.34—Use of the Grand Composite Curve for multiple utilities targeting

55

Normally we want to maximise the use of the least expensive utilities. This in turn usually means that we want to maximise the use of the coldest hot utility and the hottest cold utility. The above-the-pinch graph from Figure 2.32 (b) is reproduced in Figure 2.34 (a). Suppose the lowest level steam available on the site has a condensing temperature of 165°C, and that ΔT_{min} for the network problem is 10°C. When plotting this level on the grand composite curve, it must be plotted at 165°C − 1/2 ΔT_{min}, i.e. 160°C. This is because all temperatures in the grand composite are interval boundary temperatures, i.e. 5°C below hot stream temperatures and 5°C above cold stream temperatures. The utility line numbered 1 in Figure 2.34 (a) at 160°C represents the steam at 165°. It shows that the maximum amount of steam that can be supplied at this level is 600 kW. the other 400 kW must be supplied from high pressure steam. Suppose now we ask the question, "if I want to supply this 600 kW at a lower steam pressure, what is the minimum allowed?" The answer is given by the utility line numbered 2, showing that 600 kW can just be supplied at an interval boundary temperature of 137.5°C, i.e. a steam condensing temperature of 142.5°C. If the low pressure steam level is lowered any further, then the amount that can be supplied from it must fall. So, for example, if the steam condensing temperature is 135°C (utility line 3), then only 480 kW can be supplied at this level and high pressure steam must be increased from 400 to 580 kW.

Figure 2.34 (b) shows similar principles applied to cold utilities placed below the pinch (only now on a different example). It is desired to raise steam for the low pressure main at 140°C interval boundary temperature, starting from boiler feed water at 100°C interval boundary temperature. We want to find what is the maximum amount of steam-raising possible against the given process source profile. Knowing the shape of the steam-raising pre-heat/evaporation from physical properties, it can be constructed on the graph so that it just touches the process source profile at some point. The profile that just touches gives the maximum steam-raising possible, as shown in Figure 2.34 (b). The rest of the utility cooling must be done by lower temperature utility, in this case cooling water. Note that the point of closest approach for the steam-raising line is not necessarily at the saturation point. Note too that the variable temperature cooling water is truly represented.

2.2.5.2. Designing for many utilities—The Grand Composite Curve constructed from Problem Table analysis can be used as a design tool by the engineer wanting to specify utilities. Using the objective of maximising the use of the least expensive utilities, the shape of the grand composite often dictates the most appropriate choice of levels and loads. This is illustrated in Figure 2.35 (a). However, in many processes there is no "natural" choice of utility levels and loads, as illustrated in Figure 2.35 (b). Several choices and combinations of choices are possible in the low temperature region of this problem. Using many levels in this region clearly reduces driving forces between utilities and process, and so notionally reduces the operating cost for steam raising *etc.* However, each extra level "costs" complexity in design. The designer needs to balance the gain in running cost against the increased capital cost brought about by increasing the number of levels. The complexity implications will be described shortly.

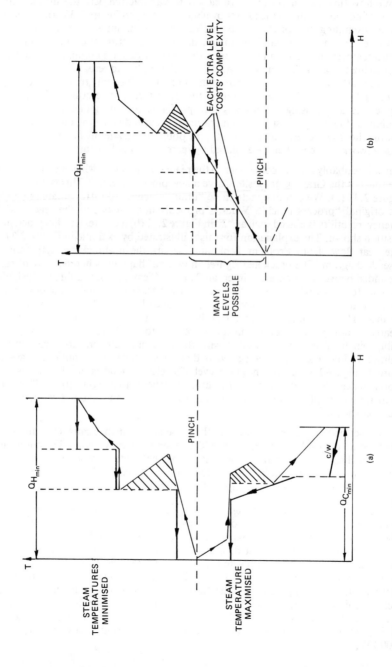

Figure 2.35—Use of the Grand Composite Curve for fixing utility levels

57

We now turn our attention more closely to variable temperature utilities, in the shape of hot oil circuits of the type shown schematically in Figure 2.36 (a). Costs in the circuit excluding the process duty exchangers are minimised by minimising oil flowrate. This means maximising the oil supply temperature and minimising the oil return temperature. Costs in the process duty exchangers are minimised by maximising driving force, *i.e.* by maximising both the oil supply and return temperatures. For the optimum total system we therefore maximise the oil supply temperature (usually dictated by oil stability) and optimise the oil return temperature. This principle applied to design using the process sink profile is shown in Figure 2.36 (b). The minimum flow for given supply temperature is predicted when the oil cooling profile just touches the process sink profile at some point. The trade-off between circuit cost and network cost can then be explored by increasing this flowrate.

It will probably have become clear to the reader by now that whenever a utility profile touches the Grand Composite Curve a new pinch is created. This is illustrated in Figure 2.37 (a). We can designate these "utility pinches" to differentiate them from the original "process pinch". What is probably not obvious to the reader is that whenever a utility is placed such as "2" in Figure 2.37 (b) it creates a utility pinch in the position shown. The explanation for this is illustrated by example in Figure 2.37 (c). The heat flows from Problem Table analysis are the un-bracketed figures in the cascade diagram. The bracketed figures show what happens when 600 kW of steam are added between temperature intervals 2 and 3. Clearly the hot utility to the highest level has to go down by 600 kW to 400 kW. However, *all* the heat flows at or above the point of addition of the 600 kW must be reduced by 600 kW, by simple enthalpy balance. Thus the flow between intervals 1 and 2 is reduced to zero, *i.e.* a pinch is created. Adding utility levels can therefore lead to the formation of two types of utility pinch. These are associated with the *load* and temperature *level* limits on utilities. In Figure 2.37 (a), for a given load on level 2, the level cannot be reduced any more. In Figure 2.37 (b), for the given level of level 2, the load cannot be increased any more. Hence it is possible to create two utility pinches with just one utility. This would happen for example if level 2 in Figure 2.37 (b) kept the same load but was reduced to the level shown.

As described in sub-section 2.2.4.2, the presence of pinches causes the number of units required to solve the design problem to increase above u_{MIN}. The addition of utility pinches over and above the process pinch can therefore cause a marked increase in the number of units required. It is often better to "relax" either load or level of the intermediate utility in order to simplify the network.

2.2.5.3. Multiple utilities design example—Design with multiple utilities will now be illustrated by example. The process stream data are given in Figure 2.38 (a). The reader may confirm that if this data set is analysed by the Problem Table method for $\Delta T_{min} = 10°C$, then the cascade diagram shown in Figure 2.38 (b) is obtained. From this diagram, the Grand Composite Curve shown in Figure 2.38 (c) can be plotted. Suppose now we want to raise VLP (very low pressure) steam at 110°C from feed water at the same temperature. If the global ΔT_{min} of 10°C also applies to utilities, then the cold utility line for the VLP must be drawn under the process source profile at 115°C. This is shown in Figure 2.38 (c), where it is also shown that the maximum VLP steam-raising possible is 460 kW out of a total cold utility requirement of 2000 kW.

58

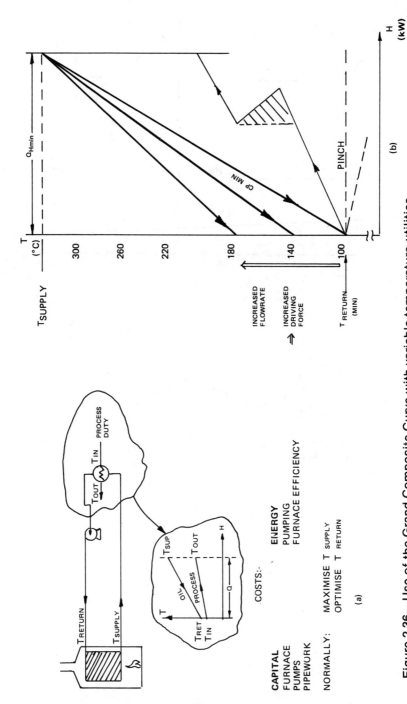

Figure 2.36—Use of the Grand Composite Curve with variable-temperature utilities

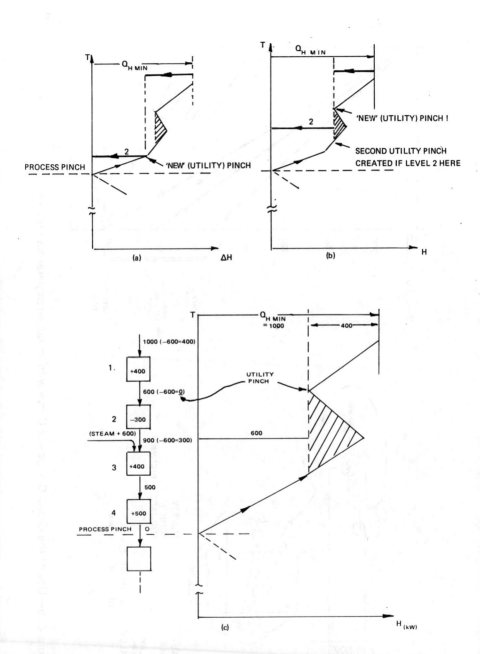

Figure 2.37— Creation of "utility pinches"

60

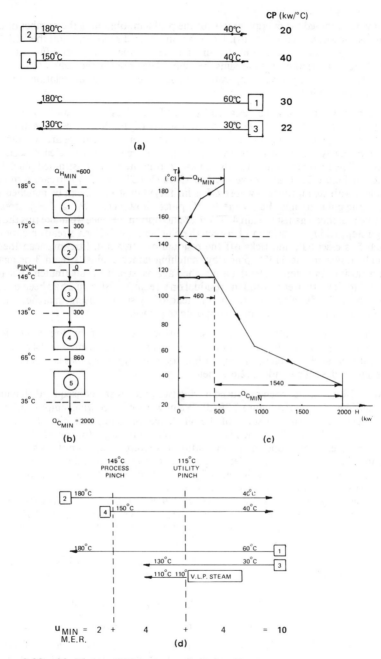

Figure 2.38—Multiple utilities example: targeting

Having obtained the energy targets for the problem, obtaining the number of units target is shown in Figure 2.38 (d). Note that in the grid representation the VLP cold utility has been drawn as a cold stream. In general, wherever utilities fall within the temperature range of process streams they should be drawn in the same way as process streams. This is because temperature cannot be ignored in relation to these intermediate utilities.

The network can now be designed by the "Pinch Design Method", yielding the solution shown in Figure 2.39 (a). The philosophy of the Pinch Design Method is to start the design at the pinch and move away. However, where there are two pinches, designing away from each into the region in between them can clearly lead to a "clash". The recommendation is, design away from the most constrained pinch first. Hence in Figure 2.39 (a), above the 115°C pinch ($CP_H \leqslant CP_C$) stream 4 must be matched with steam-raising (whose CP is infinity) but stream 2 can be matched with any of the cold streams. Below the 145°C pinch, however, ($CP_H \geqslant CP_C$), stream 1 must be matched against stream 4. There are no options. Hence we make this the first design step, ticking off stream 1 between the pinches (match 2). As already mentioned, match 5 is essential, and ticks off the residual on stream 4. Stream 2 can then be brought down to the 115°C pinch by matching either against stream 3 or against steam-raising. In Figure 2.39 (a) it is matched against steam-raising (match 4), leaving match 3 to tick off stream 3 and the residual on stream 2. If stream 2 had been brought down to the 115°C pinch by stream 3 instead of by steam-raising, matches 3 and 4 would simply have ended up in the opposite sequence.

Design above the 145°C pinch is straightforwardly done by the pinch design method. Design below the 115°C pinch, however, requires streams 3 and 4 each to be split into parallel branches. An algorithm for deciding when "stream splitting" is necessary is described under 2.2.6.2. below.

An MER design has been produced having two pinches and consequently having 10 units as demonstrated in Figure 2.38 (d). We now want to see what simplification can be achieved by sacrificing some of the VLP steam-raising, switching it to cooling water. If the 115°C utility pinch is removed, the units target for the whole problem becomes seven. The scope for simplification is therefore three units. Also, the two stream splits should not be needed. Breaking the loop shown in dotted line in Figure 2.39 (a) leads to the topology shown in Figure 2.39 (b). Note that both matches 4 and 7 and the two stream splits disappear by this operation. Also, matches 3 and 6 can then simply be merged to form a new match (match 9 in Figure 2.39 (c)). Hence three units are eliminated. Recalculating temperatures, matches 5 and 8 have a 4°C ΔT_{min} violation. This can be restored by shifting 160 kW of load from VLP steam-raising to cooling water, giving the design shown in Figure 2.39 (c).

Thus by "backing off" to 65% of the maximum possible VLP steam-raising, a design is produced that saves three units and two stream splits.

Summarising this section on multiple utilities, the designer should:

- Plot the Grand Composite Curve from a Problem Table analysis.

- Choose utilities based on information from the graph, and knowledge of process constraints (*e.g.* works mains pressures, site steam availability, safety, *etc.*).

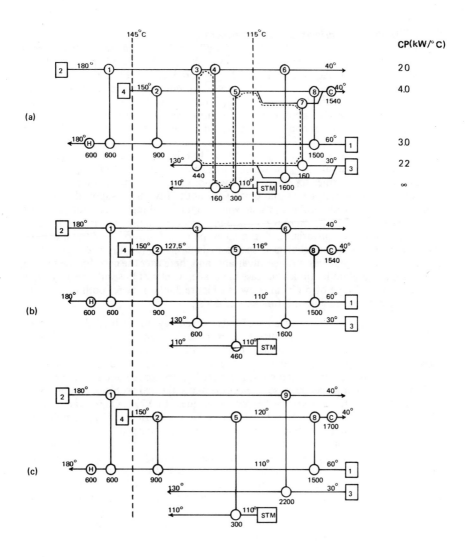

Figure 2.39—Multiple utilities example: design

- Produce an MER design, with utility pinches present.
- Relax the design, sacrificing either intermediate utility load or level to gain simplification.

2.2.6 Applying the principles—some further aspects

2.2.6.1 Is there a trade-off?—"Threshold" problems

In discussing energy targeting by composite curves, a system was shown (in Figure 2.13) where both hot and cold utilities were required. Even if ΔT_{min} were reduced to zero in this system, and hence capital cost increased to infinity, the need for both hot and cold utilities would remain. This is not true of all problems, as illustrated in Figures 2.40 (a) to (c). In Figure 2 40 (a) a pair of composite curves are shown relative to each other on the T/H plot such that both steam and cooling water are required. If however, the value of ΔT_{min} is reduced by shifting the curves together, there comes a point when one of the utilities disappears as shown in Figure 2.40 (b). In this case, the need for cooling disappeared. The value of ΔT_{min} at which this occurs is termed "$\Delta T_{threshold}$". If the curves are shifted further together, this does not cause a further change in utilities requirements. Rather, it means that part of the hot utility can be supplied at the low temperature end of the problem, as shown in Figure 2.40 (c). Hence, for values of ΔT_{min} less than $\Delta T_{threshold}$ *there is no trade-off between energy and capital,* because the utilities usages are invariant. Plotting a graph of ΔT_{min} *versus* utilities usage for this "threshold" type of problem is shown in Figure 2.40 (d). From $\Delta T_{min} = 0$ to $\Delta T_{min} = \Delta T_{threshold}$ the hot utility is invariant, and thereafter the need for the second utility appears and both utilities increase in parallel. Contrast this with the behaviour of the "pinched" type of problem shown in Figure 2.40 (e). Here, both utilities are always present and are always a function of ΔT_{min}.

Although there can be no trade-off between energy and capital for $\Delta T_{min} < \Delta T_{threshold}$, this does not mean that all threshold problems display no trade-off. Although for values of ΔT_{min} up to threshold, energy is invariant, it is possible to increase ΔT_{min} beyond this and for total cost (energy *plus* annualised capital) to decrease. In this case, the economic value of ΔT_{min} lies somewhere in the trade-off region and for all practical purposes the problem in a pinched problem. A simple way of testing for this situation is to use "experience values" of ΔT_{min}. One can get a *rough* idea of the economic value of ΔT_{min} from previous experience. For example, the ΔT_{min} experience value for a boiler house might be $50°C$, for a general heavy chemical process $20°C$, and for a refrigeration system $5°C$. This value of ΔT_{min} can be plotted on the utilities—ΔT_{min} graph as shown in Figure 2.41 (a). If it lies well into the "pinched" region then there will almost certainly be an energy/capital trade-off, to be explored as already described. An MER design will be produced by the Pinch Design Method. If the ΔT_{min} experience value lies near the threshold value, then there is likely to be a trade-off and the problem should be treated as a pinched problem. If the experience value falls well into the insensitive region then there is no trade-off and the problem cannot be designed by the Pinch Design Method.

In fact, the design of threshold problems of this type is usually "slack", *i.e.* it is free of the thermodynamic constraint of the pinch. This means that usually a great many designs are possible. The design will generally be determined by placing heaters or coolers for good control, applying the ticking-off rule, and by identifying essential matches at the "non-utility" end. This last point is illustrated in Figure 2.41 (b) for a problem requiring no utility cooling. Stream 2 must be brought down to its target temperature by interchange, and the only cold stream cold enough to do this is stream 6. This then is an essential match. Usually in threshold problems there is no need for

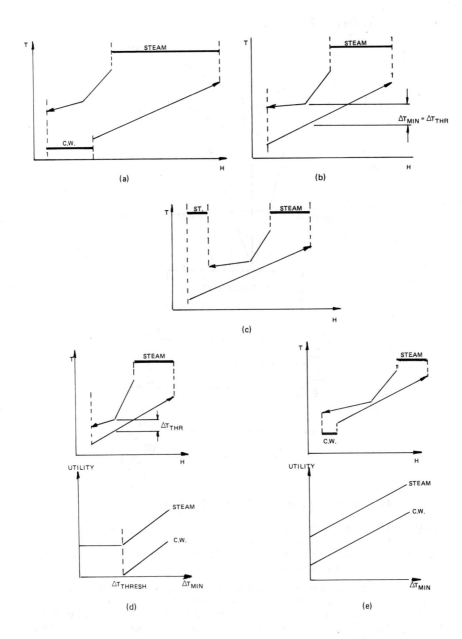

Figure 2.40—Threshold problems

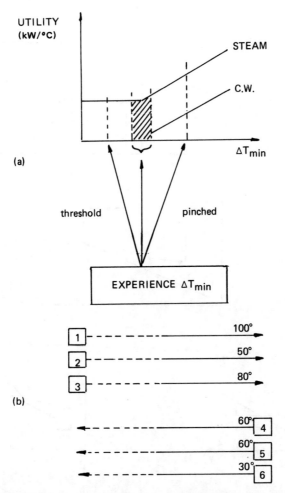

Figure 2.41—Design strategy for threshold problems

energy relaxation because the absence of a pinch means that $u_{min} = u_{min\,MER}$. In placing matches, the smallest value of ΔT is kept as large as possible, but allowed to reach some practical minimum if necessary. This value of course has nothing to do with the trade-off between energy and capital.

The reader may note that threshold problems are quite common in practice. Also, it is interesting to note that virtually all early research work in heat exchanger network design was focused—by accident—on threshold problems. This was pointed out by Linnhoff *et al* (1979). In practice, it is important for the designer to identify what type of problem he or she has prior to design, and whether or not a trade-off exists. A procedure for investigating this is given in Figure 2.42.

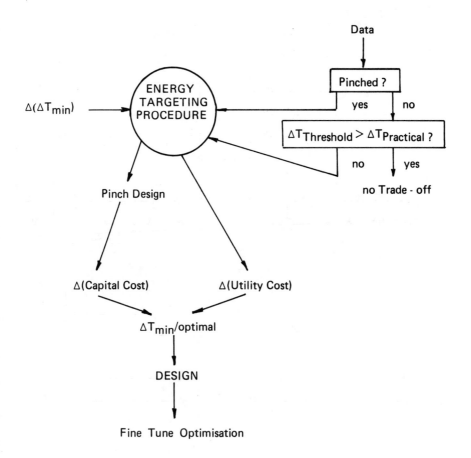

Figure 2.42—Strategy for "designing from scratch"

2.2.6.2. Stream-splitting—The principle of design at the pinch has been illustrated earlier by simple example. However, in practical more complex cases, such as the "multiple utilities" example under 2.2.5.3 above, a more comprehensive set of rules and guidelines is required. These rules and guidelines constitute a major part of the "Pinch Design Method" of Linnhoff and Hindmarsh (1982) which is the basic network design method employed in this Guide.

Consider Figure 2.43 (a), which shows an above-the-pinch stream set. For MER design, utility cooling must not be used above the pinch, which means that all hot streams must be cooled to their pinch temperature by interchange with cold streams. In Figure 2.43 (a) there are three hot streams and two cold, and so regardless of stream *CP*s, one of the hot streams cannot be cooled to pinch temperature by interchange! The *only* way out of this situation is to split a cold stream as shown in Figure 2.43 (b)

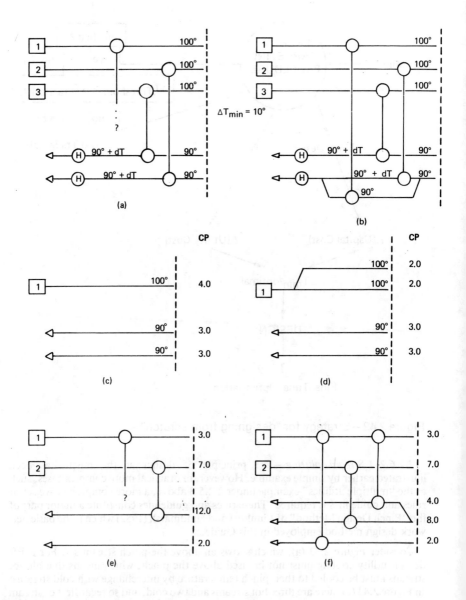

Figure 2.43—Stream-splitting at the Pinch

68

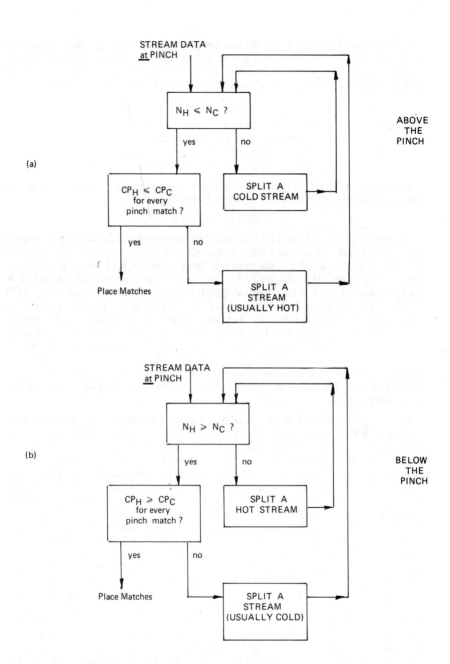

Figure 2.44—Algorithms for design at the Pinch

into two parallel branches. Now, the number of cold streams plus branches is equal to the number of hot streams and so all hot streams can now be interchanged down to pinch temperature. Hence, in addition to the CP feasibility criterion introduced earlier we have a "number count" feasibility criterion, where above the pinch,

$$N_{\text{HOT}} \leqslant N_{\text{COLD}}$$

where N_{HOT} = number of hot stream branches at the pinch (including full as well as split streams)

N_{COLD} = number of cold stream branches at the pinch (including full as well as split-streams).

Look now at Figure 2.43 (c). The number count criterion is satisfied (one hot stream against two cold streams) but the CP criterion

$$CP_{\text{HOT}} \leqslant CP_{\text{COLD}}$$

is not met for either of the possible two matches. In this example the solution is to split a hot stream as shown in Figure 2.43 (d). Usually in this type of situation the solution is to split a hot stream, but sometimes it is better to split a cold stream as shown in Figures 2.43 (e) and (f). In Figure 2.43 (e) the number count criterion is met, but after the hot stream of $CP = 7.0$ is matched against the only cold stream large enough ($CP = 12.0$), the remaining hot stream of $CP = 3.0$ cannot be matched against the remaining cold stream of $CP = 2.0$. If a hot stream were now to be split, the number count criterion would not then be satisfied and a cold stream would then have to be split as well! It is better to split the large cold stream from the outset as shown in Figure 2.43 (f), producing a solution with only one split. Step-by-step procedures for finding stream splits are given for above and below the pinch in Figures 2.44 (a) and (b) respectively. The below-the-pinch criteria are the "mirror image" of those for above the pinch.

The procedure will now be illustrated by example. The stream data above the pinch are shown in Figure 2.45 (a), and the CP data are listed in Figure 2.45 (b) in what we shall call the "CP-table". Hot stream CPs are listed in the column on the left and cold-stream CPs in the column on the right, and the relevant CP criterion noted in the box over the table. There are two possible ways of putting in the two required pinch matches, shown at the top of figure 2.45 (c). In both of these, the match with the hot stream of $CP = 5.0$ is infeasible, hence we must split this stream into branches $CP = X$ and $CP = 5.0 - X$ as shown in the bottom table in Figure 2.45 (c). Now, $CP_{\text{H}} = X$ or $5.0 - X$ can be matched with $CP_{\text{C}} = 4.0$, as shown. However, one of the split branches has no partner, $i.e.$ the number count criterion has failed and a cold stream must be split. Either $CP_{\text{C}} = 4.0$ or $CP_{\text{C}} = 3.0$ could be split, and Figure 2.46 (a) shows $CP_{\text{C}} = 3.0$ split into branches Y and $3.0 - Y$. To find initial values for X and Y it is recommended that all matches except for one are set for CP equality. Thus in Figure 2.46 (b), X is set equal to 4.0 and Y set equal to 1.0, leaving all the available net CP difference (i.e. $\Sigma CP_{\text{C}} - \Sigma CP_{\text{H}}$) concentrated in one match. The procedure quickly identifies a set of feasible limiting values. Starting from this set, it is then easy to redistribute the available CP difference amongst the chain of matches, for example as shown in Figure 2.46 (c). This design is shown in the grid in Figure 2.46 (d). The way in which the branch CPs are distributed is often dictated by the loads required on individual matches by the "ticking-off" rule.

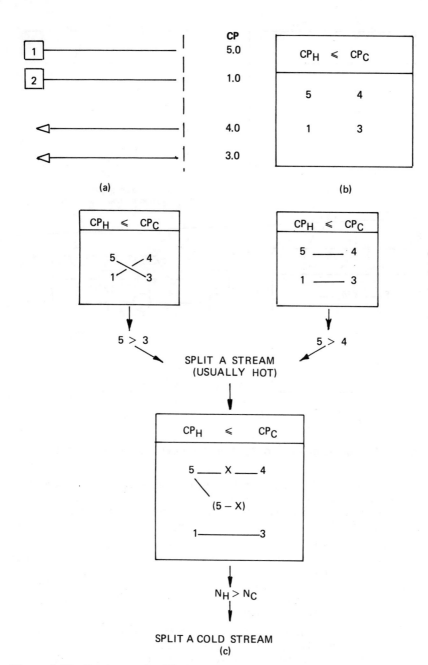

Figure 2.45—Design at the Pinch: use of the "CP-table"

71

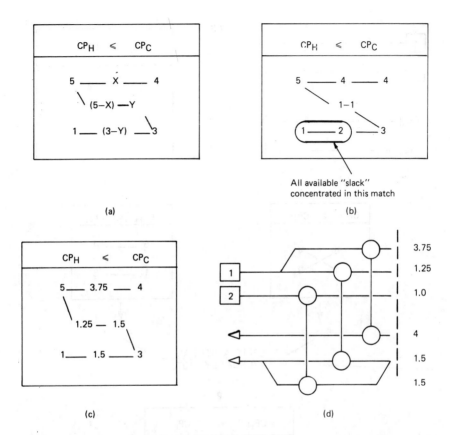

Figure 2.46—Determination of split branch flows using the *CP*-table

The procedure described in this section begs the question "is it always possible to find a solution to the pinch design problem?" The answer to this question is "yes", as can be appreciated by remembering the composite curves. Above the pinch $\Sigma CP_H < \Sigma CP_C$, and below the pinch $\Sigma CP_H > \Sigma CP_C$, are always true.

Finally, it will be clear to the reader that stream splitting at the pinch will commonly be required to produce an MER design. In some cases this may not be a desirable feature. However, stream splits can be evolved out of the design by energy relaxation, in a manner similar to the energy relaxation for reduction in number of units. The two go very much hand-in-hand, as has been seen in the multiple utilities design example under 2.2.5.3, and as will be shown in the Revamp Studies section, 2.2.6.5.

2.2.6.3 Design away from the pinch—It has been shown that if for each design decision at the pinch the designer maximises match loads to tick-off streams or residuals, then a u_{min} solution results. However, in many problems it is not possible to do this in the simple way illustrated (Figures 2.20 (c) and 2.21 (b)).

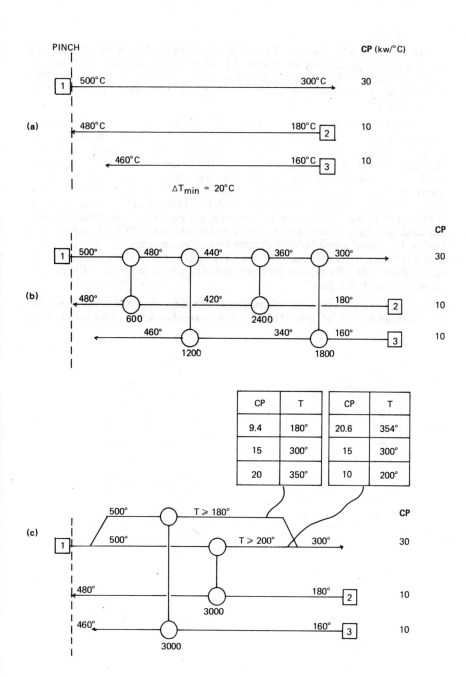

Figure 2.47—Design away from the Pinch

73

Consider the example shown in Figure 2.47 (a). Analysis of the stream data shows a pinch at the supply temperature of stream 1 and the target temperature of stream 2 and hot and cold utility requirements both of zero. The design problem is therefore entirely "below the pinch", with only one pinch match possible, *i.e.* that between streams 1 and 2.

This is a feasible match ($CP_{HOT} \geqslant CP_{COLD}$), but if its load is maximised to tick off stream 2 (a load of 3000 units), stream 1 is cooled to 400°C. This is not then hot enough to bring stream 3 up to its target temperature of 460°C. Since heating below the pinch is not allowed for an MER solution, the design step of ticking off stream 2 would lead to a design that failed to reach the energy target. An alternative strategy is shown in Figure 2.47 (b). The load on the pinch match is limited to 600 kW so that stream 1 remains just hot enough (at 480°C) to bring stream 3 up to its target temperature. However, the next match (between streams 1 and 3) also cannot be maximised in load, because now stream 2 has to be brought up to 420°C by stream 1. The load on the second match between streams 1 and 2 has to be limited, allowing a final match (between streams 1 and 3) to finish the design. This phenomenon of stream temperatures and *CP*s causing repeated matching of the same pair of streams is known as "cyclic matching".

Cyclic matching always leads to structures containing loops and hence more than the minimum number of units. The only way to avoid cyclic matching is to employ

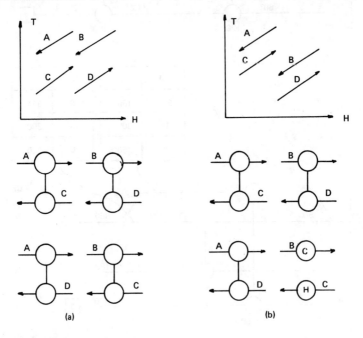

(a)   (b)

Figure 2.48—Match constraints

74

stream splitting away from the pinch. In Figure 2.47 (c) the heavy stream, stream 1, is split into two parallel branches, and each branch matched separately to a cold stream. Because this technique "slims down" the heavy hot stream it prevents the phenomenon of repeated pinching of individual matches. Hence the two matches can now be maximised to tick off the two cold streams without running into temperature problems. A u_{min} design results. Notice too that the stream split design gives an element of flexibility to the network. The split stream branch flowrates can be chosen within limits dictated by the cold stream supply temperatures. Thus if the branch matched against stream 3 is cooled to 180°C (the minimum allowed) it will have a CP of 9.4 and by mass balance the CP of the other branch will be 20.6. A CP of 20.6 in the branch matched against stream 2 leads to an outlet temperature on this branch of 354°C which is much higher than the minimum allowed (200°C). The same argument can be applied to define the other set of limits based on stream 2 supply temperature. The branch matched against stream 2 then has a CP of 20 and an outlet temperature of 350°C. The CP of the branch matched against stream 3 may therefore vary between 9.4 and 20 with the parallel limits on the other branch being 20.6 and 10. These results are summarised in Figure 2.47 (c), along with the results for equal branch flows. This type of flexibility is normally available in stream split designs and can be very useful.

To summarise, whenever the designer runs into trouble in applying the ticking-off rule, he should attempt to find a stream split design before resorting to cyclic matching. However, stream splitting does add complexity to networks as well as flexibility, hence if a non-stream-split, u_{min} solution can be found, it will normally be preferable to a stream-split solution. Note that stream splitting cannot reduce the number of units below the target value.

An example of a safe, operable, and flexible stream-split design is given in Section 4.2 of this Guide.

2.2.6.4 Constraints—Designers are always faced with many more constraints than purely thermodynamic ones when designing heat exchanger networks. Two important ones are considered in this section.

Firstly, consider the problem of "forbidden matches", as illustrated by a simple example in Figure 2.48. There are many reasons why a designer might want to forbid a match between any given pair of streams, for example corrosion and safety problems, long pipe runs required, or controllability. Imposing a forbidden match on a design might or might not affect the possible energy recovery of the network. At the top of Figures 2.48 (a) and (b) are shown four streams, two hot (A and B) and two cold (C and D). In Figure 2.48 (a) it is clear that because the relative temperatures allow, either of A or B may interchange with either of C or D. Forbidding a match between, say, A and C does not impair the chances of producing an MER design. However, in Figure 2.48 (b) it can be seen that B is not hot enough to exchange with C, and so a match between A and C is essential if an MER design is to be produced. The consequence of forbidding the A—C match is therefore an increase in utilities as shown at the bottom of Figure 2.48 (b). In general heat exchanger networks, the problem of deciding whether or not a forbidden match constraint will affect the energy target, and if so by how much, is a difficult one. However, the linear programming method of Cerda *et al* (1982) rigorously solves this problem. It is a method well suited to programming on

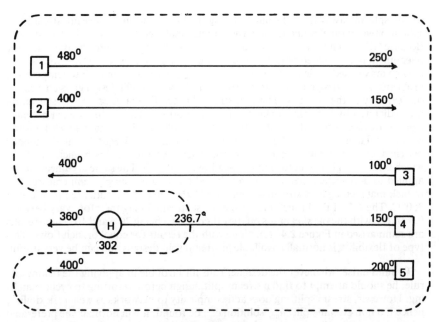

Figure 2.49—Remaining problem analysis

the computer and can be made to run very fast even on large problems. Hence with the constrained problem it is possible to retain the rigorous energy targeting element. Further, it allows the designer to define precisely what energy penalty he is paying for the constraint, and so the cost incentive for overcoming it (*e.g.* by the use of a different, possibly more expensive, mechanical design). When it comes to a design method for forbidden-match problems, an equivalent of the pinch design method has not yet been developed. At the moment, the best way to approach the problem is probably to produce an "unconstrained" MER design by the pinch design method, and then to modify it in the light of the constraint and the modified energy target. This theme will be developed further in considering "imposed matches", below, and in the later subsection on Revamp Studies, 2.2.6.5.

Secondly we look at the constraint of imposed matches. For reasons of operability (*e.g.* start-up and control), layout, and in order to re-use existing units in "revamps", the designer may want to include a certain match in his design. Suppose in the example problem shown in Figure 2.49 the designer requires a heater on stream 4 for start-up and control reasons. Analysis of the data shows a total utility heating requirement of 302 units and no cooling requirement. In order to meet simultaneously the control objective and the requirement for minimum number of units, the designer would like to place the whole heating duty on stream 4. The question is, does this design step prejudice his chances of achieving an MER design? The way to test for this is to analyse the "remaining problem" indicated by the dotted line in Figure 2.49. That is, apply the energy targeting procedure to streams 1, 2, 3 and 5, and the residual of stream 4 after placement of the heater. Applying the procedure, two results are possible. Either the remaining problem will require no utility heating, in which case the heater placement does not prejudice MER design, or the remaining problem will require heating X and cooling X, in which case the full heater load cannot be placed on stream 4 for MER design to be achievable.

Remaining problem analysis is a powerful tool for checking design steps, and can be used after the placement of exchangers as well as heaters and coolers.

In summary for the three sub-sections on design methods (2.2.6.2, 3, 4) the recommended procedure for combining them is:

● Use the Pinch Design Method to place matches at the pinch.

● Place utility heaters and coolers for operability, if necessary using a remaining problem analysis as a check.

● Fill in the rest of the design, if necessary using stream splitting.

2.2.6.5. Revamp studies—Increasingly, the chemical engineer is faced with the problem of finding ways of improving the energy efficiency of existing production units. On processes designed prior to the oil crisis it is usually quite easy to find revamp schemes that save energy. The problem is to find the best ones. The method described in this section presents a logical approach.

Figures 2.50 (a) and (b) show a rather schematic representation of the population of feasible network designs against energy recovery. The population is sparse at maximum energy recovery, but increases, sometimes greatly, as driving forces are increased and energy recovery is relaxed. The existing design will be one of many towards the base of the "pyramid". To start with this design as the basis for finding improved designs is not a good idea. The best design will normally not be within easy range of evolutionary steps by the obvious routes (Figure 2.50 (a)). However, by using heat exchanger network synthesis to produce an MER design, a starting point at the top of the pyramid is obtained (Figure 2.50 (b)). Now, the designer gains an overview of the solution space and by means of the obvious evolutionary routes can steer back towards any part. This philosophy is illustrated by simple example in Figures 2.51 and 2.52. In Figure 2.51 (a) an existing design is shown. Applying the energy targeting method for $\Delta T_{min} = 10°C$ gives a utility heating target of 106.4×10^2 kW and a utility cooling target of 85.7×10^2 kW. In other words, there is a 46% scope for saving energy. Producing an MER design, there is only one option above the pinch and this is shown in Figure 2.51 (b). The heater and match 1 are both present in the base case design, but the match between streams 2 and 5 represents a "new" match. Below the pinch (Figure 2.51 (c)), the pinch design method requires one pinch match, *i.e.* that between streams 1 and 5, which is not present in the base case. After placing this match, there are a few options for completing the design. The philosophy of the approach here is, where there are options, *choose those options which maximise compatibility with the existing design*. This philosophy dictates the below-the-pinch design shown in Figure 2.51 (c), requiring no further new matches. Putting the above and below-the-pinch designs together gives the MER design shown in Figure 2.52 (a). It requires one stream split not present in the base case, and two new matches. To evolve this design at minimum energy sacrifice back towards the base case design, the first target is the "new" match carrying 22.1×10^2 kW of load. Eliminating this

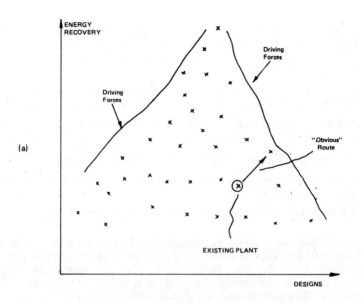

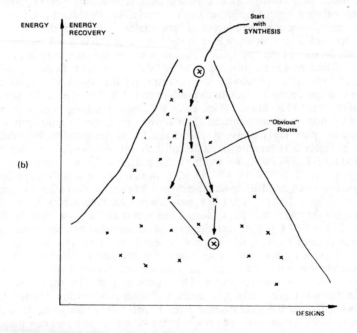

Figure 2.50—Design strategy for "revamps"

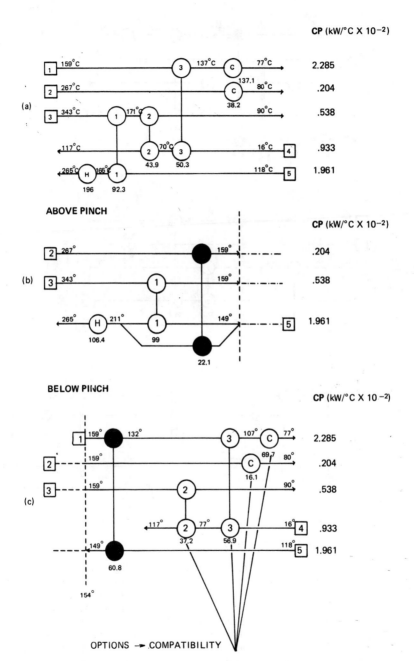

Figure 2.51—Revamp example: MER design

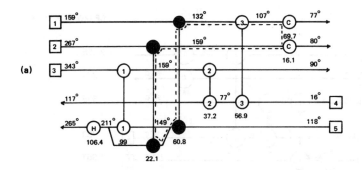

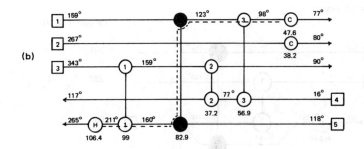

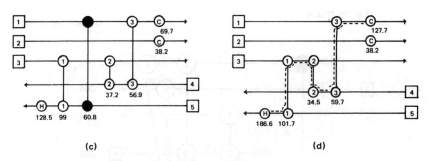

Figure 2.52—Revamp example: design evolution

match by breaking the loop picked out by dotted line in Figure 2.52 (a), the network shown in Figure 2.52 (b) is obtained. This network now has two infeasible matches, requiring energy relaxation along the path shown. If ΔT_{min} is restored, the design shown in Figure 2.52 (c) is obtained. Notice that it was necessary to relax by the full 22.1 $\times$ 10² kW lost in the eliminated match. Notice too that energy relaxation led to the elimination of the stream split. Further loop-breaking and energy relaxation with $\Delta T_{min} = 10°C$ leads to the design in Figure 2.52 (d). Notice that this is the same topological (units arrangement) design as the base case. Compared to the base case, energy has been "tightened up" along the path shown, with increase in load on matches 1 and 3 and decrease in load on match 2.

The next step is to make a crude evaluation of all the designs produced, comparing them to the base case design. At this stage, ΔT_{min} is abandoned and the effect of the network changes on the individual units is assessed. This is most simply done by "$U.A.$-analysis". By applying the well-known equation $U.A. = Q/\Delta T_{LM}$ to each unit, the effect of network changes on the total area of each unit is assessed, on the assumption that U remains constant. So, for example, in the MER design shown in Figure 2.52 (a), match 1 is 2.35 times its base case size, partly due to increased load and partly due to reduced driving force. Having done this, loads should then be shifted around loops or along paths in the networks to restore the $U.A.$ values as far as possible to their values in the base case, *but without eliminating units*. Note that full use should be made of any spare capacity a unit may have available. As a result, a table as shown in Figure 2.53 (a) can be produced, ranking the possible improvement schemes in terms of energy performance and listing the equipment modifications necessary for each. (Note, the MER topology is that shown in Figure 5.52 (a), design I topology that shown in Figure 2.52 (c), and design II topology that shown in Figure 2.52 (d)). From this table, the "best bets" are identified for further evaluation, involving detailed simulation of the network's performance.

Summarising, to find the best potential energy improvement schemes, the designer should:

- Obtain an MER design, having as great a compatibility with the base case as possible.

- Find other alternative topologies by loop breaking and energy relaxation, attempting to obtain closer compatibility.

- Perform a crude evaulation of all the alternative topologies by $U.A.$-analysis, restoring $U.A.$ values of existing units as far as possible.

- Perform detailed simulation and optimisation of the "best bets".

Finally, a word of warning: don't give up on the basis of one route only from MER design to base case! Figure 2.53 (b) illustrates that where there are options there will be more than one route.

2.2.7.   Data extraction (or, are you solving the right problem?)
In designing a process flowsheet, the process engineer normally lays down a "base case" flowsheet from a mixture of his own engineering experience and experience with similar processes. From this base case flowsheet it is very important that the stream data are "extracted" properly to ensure that an optimal heat exchanger network design results. Similarly if a revamp study is being attempted, data extraction from the process flowsheet is vitally important for best results.

2.2.7.1.   Data accuracy—In the description of energy targeting by the Problem Table method, all streams were assumed to have CPs independent of temperature. In real problems heat capacities are always dependent to some extent on temperature, and so it is important to know when the linear approximation is valid, and when it is not. Consider again the composite curve shifting shown in Figure 2.13. For a defined

CRUDE ECONOMICS – SUMMARY

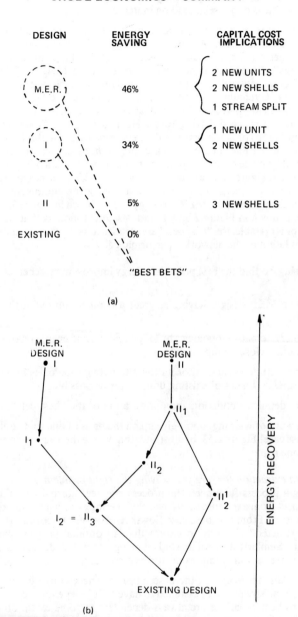

Figure 2.53—Ranking of evolved revamp schemes

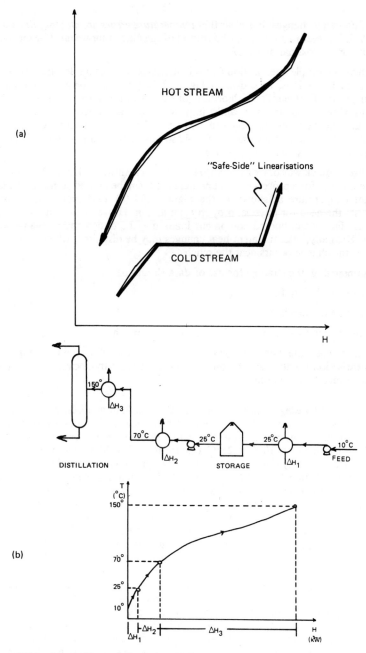

(a)

HOT STREAM

"Safe-Side" Linearisations

COLD STREAM

H

150°

ΔH_3

70°

ΔH_2

25°   25°   25°
                                              10°C
                                              FEED

DISTILLATION   STORAGE

(b)

T
(°C)
150°

70°

25°
10°

ΔH_1 ΔH_2 ΔH_3 H
                                                       (kW)

Figure 2.54—Modelling of flowsheet streams

83

set of enthalpy changes, it is clear that *temperature errors are most significant at the pinch*. It is therefore in the neighbourhood of the pinch that we must be most careful about the approximation of *CP*.

Where the simple linearisation *CP* = constant is not acceptable, streams should be linearised in sections, as shown in Figure 2.54 (a). This operation maintains the validity of the Problem Table algorithm (linear sections being handled in just the same way as whole linear streams), whilst improving accuracy where necessary. In Figure 2.54 (a), "safe side" linearisation is shown, *i.e.* the hot stream linearisation always being on the cold side, and *vice versa* for the cold stream. This ensures that predicted energy targets can always be met in practice.

A useful quick way of extracting flowsheet data is shown in Figure 2.54 (b). Where a heat balance for a process exists, unit heat loads can be plotted cumulatively against stream temperature as shown in the Figure 2.54 (b) T/H plot. This completely bypasses the need for physical property simulation. However there are two possible pitfalls. Firstly the quoted equipment loads may be design rather than operating loads. Secondly, internal latent heat changes may be disguised, which might lead to error if the change occurs near the pinch.

Summarising, the strategy for use of data should be:

● Use rough data first.

● Locate the pinch region.

● Use better data in the neighbourhood of the pinch.

Note that, when latent heat changes occur, the dew and bubble points should be fixed as linearisation points from the outset. This is because the location of the pinch is often defined by such points.

2.2.7.2. *Choosing streams*—Consider the distillation pre-heat train shown in Figure 2.54 (b). Faced with this sytem on a flowsheet, the question is, how many streams should be used to represent it? Suppose three streams are taken to represent the feed, *i.e.* one from 10°C to 25°C, one from 25°C to 70°C, and one from 70°C to 150°C. This is clearly not a good idea because when it came to design we would find each stream section "perfectly" matched to its original partner, *i.e.* we would be highly likely to generate the original flowsheet! Suppose two streams were defined, one from 10°C to the storage temperature of 25°C, and one from 25°C to the feed target temperature of 150°C. Now we stand a chance of finding different matches and improving the design. However, the storage temperature of 25°C is probably not critical. If the feed is represented by one stream running right through from 10°C to 150°C, the chances of finding an improved design are greater still. The storage temperature can be fixed at the "natural" break point between two matches. In general, the designer should in the first instance decide which supply and target temperatures he is going to define as "hard", and which as "soft". He will then pro-duce a base case heat exchanger network design, after which it may become apparent that by making changes to temperatures originally classified as hard, further improvements could be made. These decisions are clearly entirely dependent on the process technology and to some extent on the designer's experience. However, the

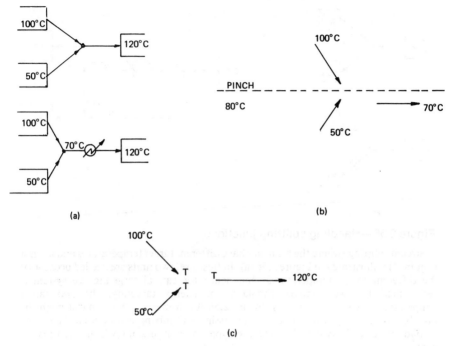

Figure 2.55—Handling mixing junctions

designer should always be on the lookout for opportunities of improving his networks by modifying the base data.

A further complication in defining streams arises when mixing and splitting junctions are involved. The top drawing of Figure 2.55 (a) shows schematically two cold streams leaving separate units at different supply temperatures, mixing and then requiring heating to a common target temperature. In terms of capital targeting, the system is really only one stream as shown in the lower drawing of Figure 2.55 (a), because it can be satisfied by only one unit. However, Figure 2.55 (b) shows what may happen if the system is regarded as only one stream for energy targeting. If the mixing temperature lies below pinch temperature, then the "cooling ability" of the cold stream below the pinch is degraded. More heat must therefore be put to utility cooling, and by enthalpy balance, heat must be transferred across the pinch increasing hot utility usage. To ensure the best energy performance at the targeting stage, the mixing must be assumed isothermal, as shown in Figure 2.55 (c). If $T = 120°C$ then the system is regarded as two streams. If $T < 120°C$, then it is three streams. Hence in stream mixing, the data for units targeting is incompatible with the data for energy targeting. However, this should not cause confusion at the design stage if the above principles are thoroughly understood. It merely means that the designer might require one more unit than minimum if non-isothermal mixing cannot be allowed in an MER design.

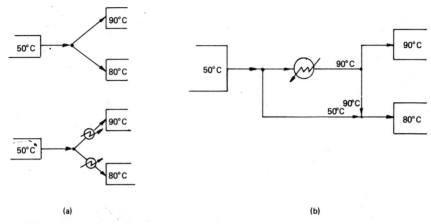

(a)   (b)

Figure 2.56—Handling splitting junctions

Stream splitting (where the branches have different target temperatures and are not re-joined) is illustrated in Figure 2.56 (a). In this case, two units are needed because of the different target temperatures, and hence for capital targeting the system is represented by two streams. Similarly for energy targeting, the two target temperatures means two streams. Figure 2.56 (b), however, shows that it might be possible to get away with one unit if bypassing and mixing can be used. Here, the second unit effectively is replaced by the bypass mixing junction which performs the heat transfer job.

Summarising this section on choosing streams, the designer should:

● Avoid over-specification of the problem.

● Look for the possibility of improving the system by changing process conditions.

● Avoid non-isothermal mixing at the energy targeting phase.

To ensure best energy performance, hot streams should be specified as hot as possible, and cold streams as cold as possible. A good slogan to remember is "keep your hot streams hot and your cold streams cold"!

2.2.8. Design strategy
The concepts and procedures described in this section of the Guide are now pulled together into an overall design strategy. This strategy is shown in outline form in Figure 2.57. It is divided into four main sections: definition of data, target setting, design and optimisation. The figure shows these four main sections further divided, with each boxed item amplified in Table 2.3. The following are important points to note:

● The "recycles" back to base case data from problem definition and from design are very important. The designer should "use" the characteristics that he discovers about the problem to help him decide how to change the problem! In other words, the heat exchanger network targeting and design procedures are very powerful tools for helping the designer evolve better *total flowsheets*.

Table 2.3

		Guide sub-section reference	Figure numbers	Notes
Data	*Data extraction*	2.2.3.2, 2.2.7	2.5, 2.6, 2.13, 2.18, 2.54–2.56	Understand flowsheet. Get stream data, and note flexibilities. Identify constraints. Check heat balance.
	Problem type, choosing ΔT_{min}	2.2.4.4, 2.2.6.1	2.29–31, 2.13, 2.40–42	Use the Problem Table method, rather than composite curves. Note Figure 2.42. Refine data where necessary.
Targets	*Energy targeting*	2.2.1, 2.2.2, 2.2.5.1, 2.2.5.2	2.5–17, 2.32–36	Use the Problem Table method, rather than composite curves. Don't blindly accept constraints.
	Capital targeting	2.2.1, 2.2.4.1, 2.2.4.2, 2.2.6.5, 2.2.7.2	2.5, 6, 20, 2.22–25, 2.53–2.56	Remember, $u_{min, MER} > u_{min}$. Use flowsheet flexibilities, force subset equality.
	Check!			Don't go any further until you're sure you've got it right!
Design	MER *design, relaxation and simplification*	2.2.3, 2.2.4.3, 2.2.5.2, 2.2.5.3, 2.2.6.1–5, 2.2.7	2.17–20, 2.25–28, 2.30, 35–56	Understand your problem type (Figures 2.41 and 2.42). Eliminate complexity (units and stream splits). Don't forget data mods to force simplicity or energy-savings.
	Optimisation	2.2.4.4, 2.2.6.5	2.27–29, 2.53	Area optimisation (*i.e.* by "load-shifting") using "*UA*-analysis". Detailed simulation.

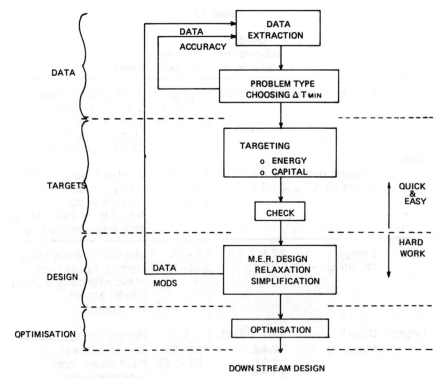

Figure 2.57—Overall design procedure

● Checking after the targeting phase is very important. Up to and including the targeting phase, the procedure is relatively quick and simple. After that it becomes more difficult, and so the designer is wise if he carefully checks his data at this point!

2.3 Heat and Power Integration

2.3.1 Introduction

So far we have seen how process heat integration can be tackled using the Problem Table analysis followed by heat exchanger network design by the Pinch Design Method. We have also seen how the information in the Problem Table can be translated into the "Grand Composite Curve" and used in the design of systems employing many utility levels. However, powerful as these methods are, they are still rather limited in scope against the background of total energy systems. Combined-heat-and-power generation is a common feature of many process industry plants and complexes, and we therefore need to consider the implications of the heat exchanger network design on the background power system.

Figure 2.58 shows a schematic energy flow diagram for a "typical" industrial site. A

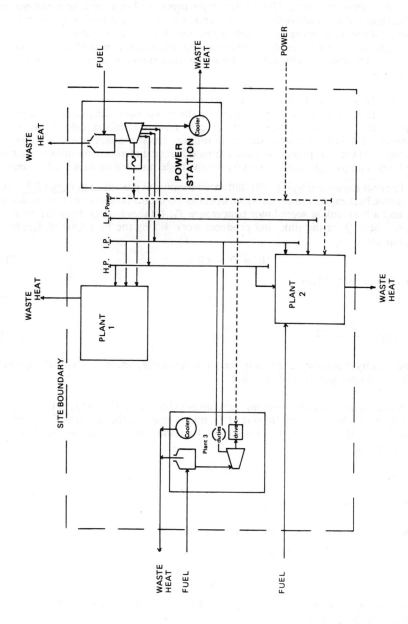

Figure 2.58—Typical site CHP system

central power station produces electrical power and steam at three levels for consumption by various processes. One of the major processes has a large heat and power requirement which is satisfied by on-plant steam-raising and direct-drive steam turbines. The total system is evidently a complex one and clearly what is needed is a thermodynamic analysis tool to enable us to check whether, overall, the system is as efficient as it can practically be. This section of the Guide describes just such a tool, and its use in context.

2.3.2. Background: heat engines and heat pumps

At the heart of all total energy systems is the power-producing (and consuming) equipment. Thermodynamically, all process power-producing devices are heat engines. Similarly refrigeration and vapour recompression systems (power-consuming) are heat pumps. A basic thermodynamic understanding of heat engines and heat pumps is therefore necessary for the understanding of integrated systems.

The heat engine concept is essentially a simple one, as illustrated in Figure 2.59 (a). A simple heat engine is a device which operates between a heat source at temperature T_1 and a heat sink at some lower temperaure T_2. It takes heat Q_1 from the source, rejects heat Q_2 to the sink, and produces work W. By the First Law of Thermodynamics:

$$W = Q_1 - Q_2 \qquad \dots\dots(1)$$

and by the Second Law

$$W \leqslant \eta_c . Q_1 \qquad \dots\dots(2)$$

where $\eta_c = \dfrac{T_1 - T_2}{T_1}$ $\qquad \dots\dots(3)$

The equality in equation 2 applies to thermodynamically ideal or "reversible" engines. All real engines fall short of this ideal.

A heat pump is simply a heat engine running in reverse, as illustrated in Figure 2.59 (b). It accepts heat Q_2 from the sink at T_2, rejects heat into the source at T_1, and consumes work W. Again, by the First Law of Thermodynamics, equation 1 applies and by the Second Law:

$$W \geqslant \eta_c . Q_1 \qquad \dots\dots(4)$$

where $\eta_c = \dfrac{T_1 - T_2}{T_1}$ as before.

Again, the equality in equation 4 applies to the reversible case.

It is usual to define overall efficiencies for real heat engines and heat pumps, according to:

$$W = \eta . Q_1 \qquad \dots\dots(6)$$

with $\eta < \eta_c$ for heat engines, and $\eta > \eta_c$ for heat pumps. Often, equation 6 is rewritten for heat pumps as:

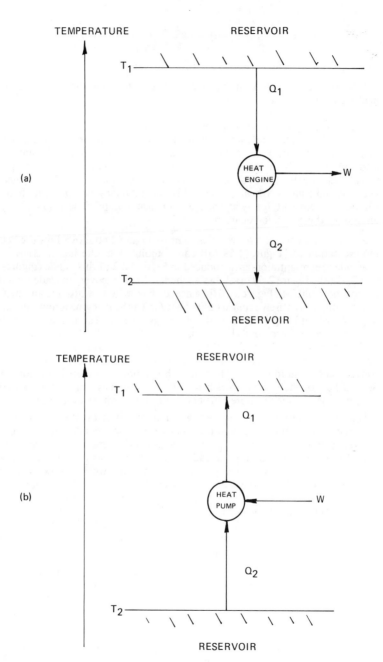

Figure 2.59—Schematic heat engines and heat pumps

$$W = \frac{Q_1}{\phi} \qquad \qquad \dots\dots(7)$$

where ϕ is the so-called "coefficient of performance".

With the above understanding, we can begin analysing the combined power and heat recovery problem.

2.3.3. Heat engines and heat pumps with processes—basic principles

In Section 2.2 of the Guide we saw how fundamentally important the pinch concept is in producing best heat exchanger networks. We now show that this same concept is also fundamentally important in the design of best combined heat and power networks. Because the pinch concept reveals the true nature of heat recovery networks as net heat sinks and net heat sources, this understanding can be combined with the source/sink concept for heat engines and heat pumps. The result is a powerful analysis of the combined problem.

2.3.3.1. Placement relative to the pinch—Figure 2.60 shows how the heat engine representation (of Figure 2.59 (a)) can combine with the heat recovery problem "cascade" representation as introduced in Section 2.2 of this Guide (Figures 2.15 (c) and 2.32 (a)). Figure 2.60 (a) shows a schematic heat recovery cascade requiring heat F_1 from hot utility. In Figure 2.60 (b) a heat engine is shown whose exhaust heat $(Q-W)$ is hot enough to entirely replace utility heat F_1. The heat engine is now consuming hot utility Q, which is equal to $F_1 + W$ (by energy balance around the heat engine). But the heat recovery problem alone consumes hot utility F_1. Hence, placing the heat engine as shown in Figure 2.60 (b) allows us to burn extra fuel equivalent to a heat quantity W in order to produce a quantity of work W. In other words, 100% efficient conversion of heat to work!* The reader should note that there is no violation of the Second Law of thermodynamics here! The 100% comes from a *comparison* of two systems, one of which is more efficient in absolute terms than the other.

Figure 2.60 (c) shows what happens when the exhaust from the heat engine is colder than the process pinch temperature, and we attempt to integrate it into the heat recovery problem. Because it is colder than pinch temperature, it cannot replace any of F_1. Consequently, it simply "cascades" through the system, eventually ending up as cooling duty at the lowest level. So in order to produce work W now, we have had to burn extra fuel Q, *i.e.* the work is produced at no more than "normal" engine efficiency. In fact, we might just as well have sent the exhaust heat direct to the ambient sink. The fundamental problem with the arrangement in Figure 2.60 (c) is that the heat engine is *transferring heat across the pinch*. The reader will recall from the discussion under Section 2.2 of this Guide that transferring heat across the pinch always leads to an energy penalty.

Finally, Figure 2.60 (d) shows what happens when the heat engine, instead of taking heat from hot utility (fuel), takes it from the surplus heat F_{n+1} available below the pinch. Now, with Q equal to F_{n+1}, by energy balance a quantity of work W is produced from waste heat W. The 100% efficiency is obtained once again.

---

*Note, "secondary" heat losses in practical systems, such as power conversion (generator) losses, mean that the conversion is not quite 100%. However, the "100%" is useful for conceptual understanding.

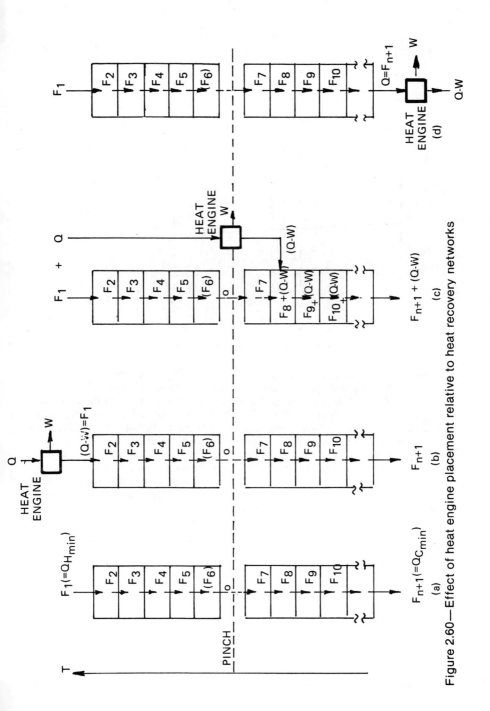

Figure 2.60—Effect of heat engine placement relative to heat recovery networks

93

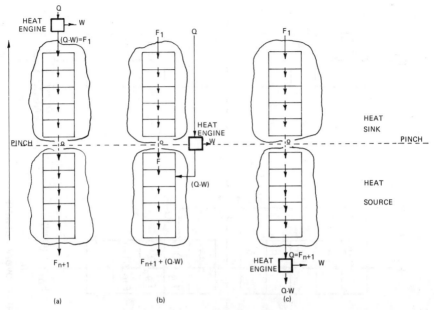

Figure 2.61—Interpretation of the "appropriate placement" result

A good conceptual understanding of what is happening can be gained by reference to Figure 2.61. The heat recovery cascade diagrams are shown bounded by enthalpy balance "envelopes" demonstrating the source/sink characteristic of the total heat recovery problem (compare Section 2.2 of the Guide, Figures 2.17 and 2.32). Thus when the heat engine rejects heat above the pinch (Figure 2.61 (a)), it is rejecting heat *into the process sink*. In so doing, it exploits the temperature difference which exists between the utility source and the process sink, producing work at high efficiency. Conversely, when the heat engine takes heat from below the pinch (Figure 2.61 (c)), it is taking heat *from the process source,* exploiting the temperature difference between the process source and the ambient sink. However, when the heat engine rejects heat below the pinch (Figure 2.61 (b)), it is rejecting heat into what is already a heat source! Consequently, no benefit from the integration can be obtained.

The above arguments essentially constitute the "Appropriate Placement" concept for heat engines of Townsend and Linnhoff (1982a). A heat engine placed either entirely above or entirely below the pinch (as in Figures 2.61 (a) and (c)) is "appropriately placed" because it confers work generation at 100% efficiency. However, a heat engine placed across the pinch (Figure 2.61 (b)) is "inappropriately placed" because it can confer work generation efficiencies no higher than the stand-alone engine.

The Appropriate Placement concept is readily extended to heat pumps. In Figure 2.62 (a), a heat pump is shown placed entirely above the pinch. It can be seen that all this heat pump succeeds in doing is replacing a quantity of hot utility W by work W, which will rarely (if ever) be a worthwhile swap. In Figure 2.62 (b) a heat pump

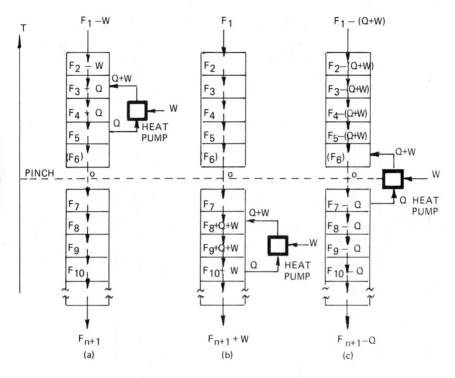

Figure 2.62—Effect of heat pump placement relative to heat recovery networks

operates entirely below the pinch, converting work W into waste heat W! Energy is saved overall only when the heat pump operates *across the pinch* as shown in Figure 2.62 (c), *i.e.* when it pumps *from the process source to the process sink*. Clearly, Appropriate Placement for heat pumps means placement across the pinch, when the "normal" efficiency is obtained. Placement on either side of the pinch is inappropriate, leading to energy wastage.

2.3.3.2. Limiting Cases—There are limits on the extent to which the 100% efficient conversion of heat to work can be exploited. These relate to engine heat *load* and heat *level*.

The load limit is illustrated in Figure 2.63 (a) for an appropriately placed heat engine above the pinch. Suppose the load on the heat engine is increased beyond the point where the quantity of exhaust heat is just equal to the net process hot utility requirement (*i.e.* beyond the point where $(Q - W) = F_1$). The incremental exhaust heat $\delta(Q - W)$ cascades down through the system, *crosses the pinch,* and eventually ends up in cooling water. Thus the incremental work δW is produced at only the stand-alone cycle efficiency of the heat engine. Hence if the power required is greater than that which can be produced at 100% efficiency the excess heat is best produced entirely independently of the heat recovery network, with the surplus exhaust heat

95

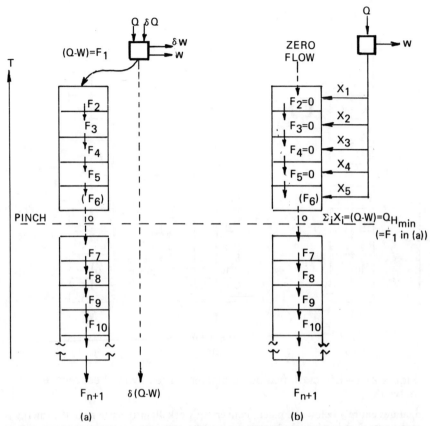

Figure 2.63—Load and level limits for appropriately placed heat engines

being rejected directly to ambient. (This maximises the efficiency of the stand-alone cycle by exploiting the lowest possible sink temperature—refer back to equations 2 and 3.) Similarly for appropriately placed heat engines below the pinch, if the heat input is increased beyond the point where it equals the net process heat surplus, heat has to be cascaded across the pinch. Incremental work generation at only stand-alone efficiency results.

The level limit is illustrated in Figure 2.63 (b) for an appropriately placed heat engine above the pinch. It should be apparent to the reader that since the process heat sink exists over a range of temperature, an appropriately placed heat engine need not supply exhaust heat all at the highest temperature (as implied by Figure 2.60 (b)). It can be supplied over a range of temperatures above the pinch. The advantage of this is that lowering average exhaust levels increases engine cycle efficiency, as can be appreciated from equations 2 and 3. But since an appropriately placed engine always produces work at 100% efficiency, the net effect is to increase the *quantity* of work available. This important (and intuitively non-obvious) conclusion is exploited when we consider total site systems under 2.3.7, below. Figure 2.63 (b) shows the limit to which exhaust levels can be reduced for an appropriately placed heat engine. It occurs when all the heat flows in the heat recovery cascade above the pinch are reduced to zero, *i.e.* all heat transfer is at minimum driving force. Note though that the total *quantity* of heat exhausted to the process sink is unchanged, *i.e.* it is equal to F_1, the "load limit". Thus the arrangement in Figure 2.63 (b) represents the maximum work that can be produced at 100% efficiency by an appropriately placed heat engine above the pinch. Exactly analogous arguments apply for the level limit for appropriately placed heat engines below the pinch, with work output at a maximum when all process heat flows are reduced to zero and the quantity of heat extracted is equal to F_{n+1}.

2.3.3.3. Use of the "Grand Composite Curve"—In Section 2.2 of this Guide we introduced a graphical representation of the process source/sink termed the "Grand Composite Curve" for use with multiple utility designs. We now show how this representation in fact helps in the design of optimal combined heat and power systems.

In Figure 2.64 (a) a cascade diagram for a particular heat recovery problem is shown, and in Figure 2.64 (b), the appropriately placed heat engine which reduces the heat flows above the pinch to zero (compare Figure 2.63 (b)). The grand composite curve for the cascade in Figure 2.64 (a) is shown in Figure 2.64 (c) (in thin line). If a heat engine exhaust is to be matched against this profile which is everywhere at the lowest possible temperature, then it would have the profile shown in Figure 2.64 (c) in heavy line. That is, everywhere exchanging heat at ΔT_{min} with the nett process sink, to the maximum extent possible. In other words, for maximum work output, the process sink and engine exhaust must have *parallel T/H profiles*. The reader may confirm that the heat quantities in Figure 2.64 (b) exhausting from the heat engine into the process over the levels shown correspond to the graphical construction in Figure 2.64 (c).

Similarly when a heat engine is appropriately placed below the pinch, it must absorb heat along a profile which runs parallel to the process source profile for work output to be maximised.

We now recall that temperatures plotted in the Grand Composite Curve are the interval boundary temperatures from the Problem Table analysis (see sub-section 2.2.2 under "Heat Exchanger Networks", Figures 2.14 and 2.15). Hence if the ΔT_{min} value for the heat recovery problem in Figure 2.64 (a) applies to the heat engine exhaust in Figures 2.64 (b) and (c), then the exhaust profile and the process sink profile in Figure 2.64 (c) should coincide (the exhaust profile has been shown slightly displaced vertically for clarity). In general though we might well want to assign a different ΔT_{min} value to exhaust/process matches than to process/process matches (because the values will be determined by different tradeoffs). This might be thought to cause problems. The Problem Table algorithm, however, can be adapted to cover this case *i.e.* where ΔT_{min} is "match-dependent" and is not just a global value. This can be achieved by assigning ΔT_{min} "contributions" to streams. Thus if liquid streams are assigned a contribution of 5°C and gas streams 10°C, then a liquid/liquid match has a ΔT_{min} of 5 + 5 = 10°C, a liquid/gas match has a ΔT_{min} of 5 + 10 = 15°C and a gas/gas match has a ΔT_{min} of 10 + 10 = 20°C. In setting up temperature intervals in the Problem Table procedure, the interval boundaries are then set at hot stream temperatures minus their ΔT_{min} contribution, rather than half the global ΔT_{min}. Similarly, boundaries are based on cold stream temperatures plus their ΔT_{min} contribution. An example stream set and ΔT_{min} contributions are shown in Figure 2.65

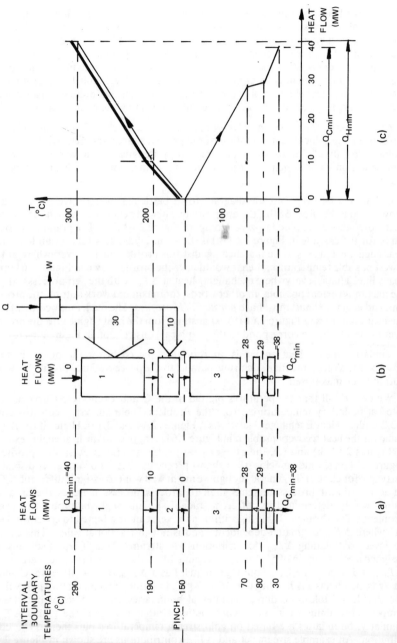

Figure 2.64—Best thermodynamic match of heat engine exhaust with the Grand Composite Curve

98

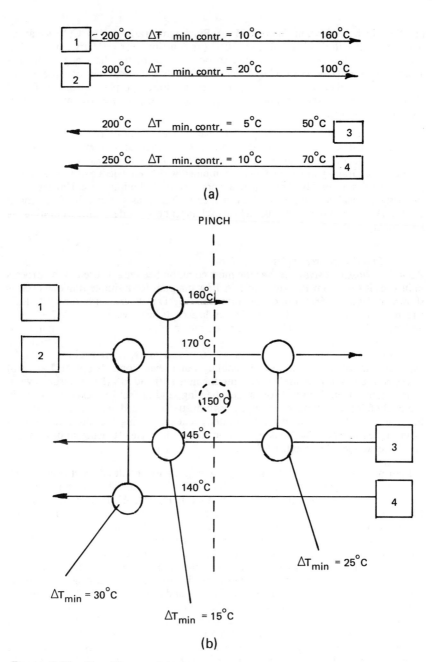

Figure 2.65—Handling match-dependent ΔT_{min}

99

(a). The pinch is shown in Figure 2.65 (b). Pinch temperature is 150°C, hence at the pinch, stream 1 (hot) is at 150°C + 10°C (its contribution) = 160°C, stream 3 at 150°C − 5°C (its contribution), and so on. In constructing heat engine profiles against the Grand Composite Curve then, it is entirely a matter of personal preference as to whether real engine profile temperatures are plotted (displaced from the Grand Composite Curve by the appropriate ΔT_{min} contribution), or adjusted (interval boundary) temperatures.

The parallel profiles for appropriately placed heat engines represent the limiting case, and so a design based on this case will rarely be practical. However, *the principle of appropriate placement remains valid in practical design situations*. What we can do is to use the Grand Composite Curve, combined with a knowledge of practical heat engine profiles, to find the best practical approach to the limiting case. Having found the best practical engine profile, we can then include it as a stream in the Problem Table analysis, in order to produce a heat exchanger network design incorporating the heat engine profile.

2.3.4. Practical power cycles

2.3.4.1. Rankine cycles—By far the most common heat engine used in the process industries is the steam Rankine cycle. A flow diagram for a simple Rankine cycle is shown in Figure 2.66 (a) and the corresponding T/S diagram for the cycle fluid (steam) is shown in Figure 2.66 (b). The cycle accepts heat over the profile 2−3−4−5 and rejects heat over 6−1. These profiles are shown in the T/H diagram in Figure 2.66 (c). A more complex Rankine cycle is shown in Figure 2.66 (d), with corresponding T/S and T/H diagrams shown in Figures 2.66 (e) and (f) respectively. Now, instead of expanding all of the steam down to cooling water level, some "stops off" at (two) intermediate "back-pressure" levels (intermediate (IP) and low (LP) pressure levels). This exhaust steam is useful for process heating, and in fact the schematic cycle in Figure 2.66 (d) is the basis of most process industry combined heat and power (CHP) installations. Notice that boiler feedwater preheating using the intermediate level exhaust steam is possible, making significant overall cycle efficiency improvements (see for example, Haywood, 1980).

Returning to our theme of matching engine exhaust profiles against process sink profiles, the steam Rankine cycle exhaust profiles available for matching are represented by 7′−8, 9′−10, (and possibly 6−1) in Figure 2.66 (f). That is, those profiles which exist *outside* any region of internal cycle heat exchange. What do these profiles look like when matched against process sinks? The answer has already been given, in section 2.2 of the Guide under the description of Multiple Utilities! Refer back to Figure 2.35 (a) of sub-section 2.2.5.2, which is reproduced here as Figure 2.67 (a). We see that the "parallel profiles" limiting case is approximated very well by choosing the three steam exhaust levels shown above the pinch. The origin of multiple utilities of differing costs in fact lies in the heat and power problem. By maximising the use of steam at minimised levels, the Rankine cycle efficiency is maximised, leading to more power being produced at 100% efficiency. Figure 2.67 (a) implies that there is a completely free choice of steam level. This will only be the case when a "dedicated" cycle is being built for the plant in question. Where levels are fixed by site mains pressures, the best that can be done is to maximise the use of LP before IP, IP

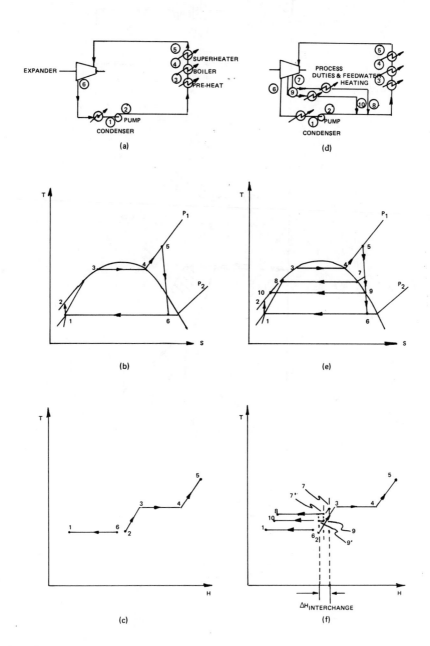

Figure 2.66—Simple and complex Rankine cycles

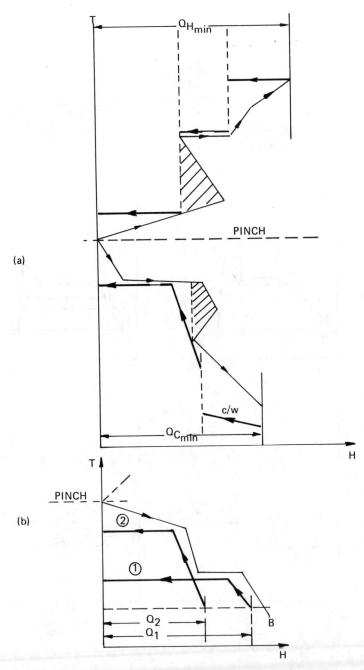

Figure 2.67—Matching Rankine cycles against the Grand Composite Curve

before HP, *etc.,* maximising the efficiency of the site power station. This will be discussed further under 2.3.7, below.

Getting the best practical match for a Rankine cycle appropriately placed below the pinch, is also illustrated in Figure 2.67 (a). The steam-raising case shown incorporates no superheating (*i.e.* the turbine accepts steam at condition 4 in Figure 2.66 (b)). Normally, however, with steam at medium to high pressure, some superheating is desirable so that it is not expanded too far into the 2-phase region.

When fitting a Rankine cycle heat-absorbing (steam-raising) profile below the pinch, an ambiguity relating to load *vs.* level considerations can exist, as illustrated in Figure 2.67 (b). Here, due to the geometry of the process source profile, the Rankine cycle profile 1 can absorb more heat than profile 2 ($Q_1 > Q_2$), but only at on average, lower temperature. Thus cycle 1 will have a lower cycle efficiency than cycle 2, but it has the compensation of carrying a higher heat load. Hence it is not possible to say *a priori* which cycle will yield the largest power output. The engineer must evaluate the two discrete cases to determine which is best.

2.3.4.2.   Gas turbine cycles—Although steam Rankine cycles are the most common type of heat engine, gas turbines are also very important and can offer very large cost and energy efficiency incentives in certain applications. Figures 2.68 (a) to (c) show the flow diagram and thermodynamic cycle for what is known as the "open cycle" gas turbine, *i.e.* a gas turbine which is open to the atmosphere at its lowest temperature point. It can be appreciated from Figure 2.68 (b) that, depending on the cycle pressure ratio P_1/P_2, point 4 may be hotter than point 2, in which case it is possible to pre-heat compressed air with hot exhaust as illustrated in Figure 2.68 (c). Improvement in cycle efficiency results. Once again, it is the profiles lying outside any region of internal cycle heat exchange that are matched against the process.

Matching open cycle gas turbine exhaust against the process sink is illustrated in Figure 2.69. The process sink profile is BA, with the pinch at B. Now, because the pinch is at above-ambient temperature, and the open cycle gas turbine is *constrained* to reject heat right down to ambient temperature, then *it is not possible for a gas turbine to be completely appropriately placed.* Some exhaust heat is constrained to be rejected below the pinch. This means that marginal benefits have to be explored, as illustrated by exhaust profiles 1 and 2 in Figure 2.69. Exhaust profile 1 implies a higher cycle efficiency than exhaust profile 2 (lower mean temperature), but wastes more heat rejected to ambient ($R_1 > R_2$).

A special type of "gas turbine" is the "off-gas" or "tail-gas" expander, which one sees recovering power from heat in pressurised process streams. The great advantage of this type of cooling is that enthalpy in the stream is converted into power at 1:1. This is because it exploits the fact that power has previously been put into the hot waste or product stream by compression elsewhere in the process. Hence it makes sense to put as much heat as possible into such streams before expansion, consistent with their required final conditions of temperature and pressure. This can be heat from fuel, but preferably it is waste heat which is available below the pinch.

2.3.4.3.   Heat pumps—A simple "mechanical" heat pump is illustrated in Figure 2.70 (a) with the thermodynamic cycle shown in Figures 2.70 (b) and (c). It can be seen that heat is absorbed and rejected essentially over latent heat profiles (with some

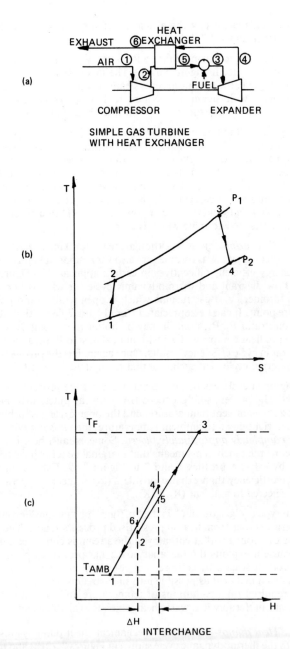

Figure 2.68—Gas turbine cycle

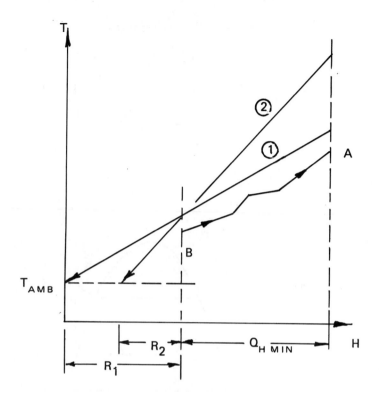

Figure 2.69—Matching gas turbine cycles against the Grand Composite Curve

desuperheating of compressed vapour on the output side). When matching against the process, it is important to remember that the load limit on *either* the process source *or* the process sink can limit the total energy saving, since Appropriate Placement for a heat pump means placement *across* the pinch.

2.3.5. Design procedure

2.3.5.1. Technology section—The Grand Composite Curve allows comparison of one heat engine type with another on the same process duty. In this connection, it is worth reiterating the important conclusion relating to the level limit. Namely, that when a heat engine is appropriately placed, the cycle efficiency affects the *quantity* of work available, but not the efficiency of work generation: Appropriate Placement always confers 100% efficiency.

Figures 2.71 (a) to (d) illustrate the relative performance of open cycle gas turbines and steam turbine Rankine cycles on two different Grand Composite Curves. In Figures 2.71 (a) and (b), the "above the pinch" process sink profile BA extends from low to high temperature. This means that the gas turbine exhaust (Figure 2.71 (a)) is most appropriate with little energy wasted by rejection below the pinch, and a high

105

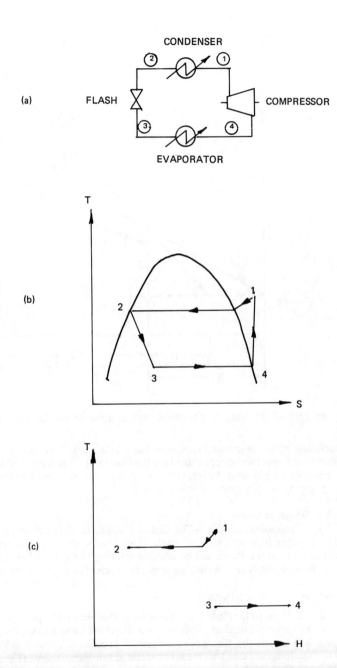

Figure 2.70—Simple heat pump cycle

106

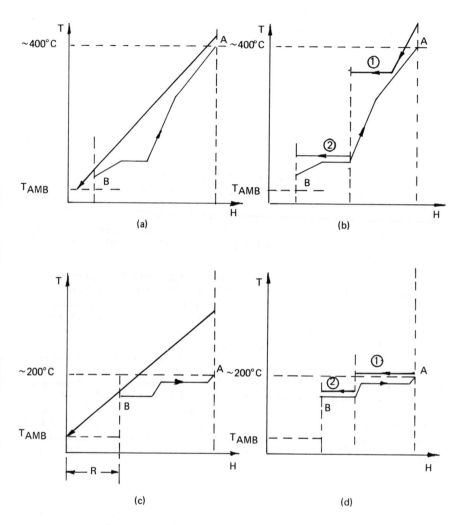

Figure 2.71—Comparison of steam and gas turbine applications

efficiency for good work output in a single cycle. Compare this with the steam turbine exhausts matched against the same process in Figure 2.71 (b). Although work can be produced at 100% efficiency, driving force losses against the sink are large. This is true even if a back-pressure turbine having two (or more) exhaust levels is used as shown in Figure 2.71 (b). Also it should be remembered that in steam cycles, the indirect heat transfer from the utility heat source to the steam in the boiler/superheater means that the upper cycle temperature is limited to a value much less than that achieved by a gas turbine. Hence driving force losses against the utility source are also higher. As a result total work output is much smaller. Even if the work required by the process itself is small and the steam cycle can provide all of it, the gas turbine system represents an opportunity of producing power for "export" to the site at a very high efficiency. This theme will be explored further in sub-section 2.3.7, below.

In Figures 2.71 (c) and (d) the process sink profile BA exists at moderate temperature. This means that the steam Rankine cycle (Figure 2.71 (d)) can achieve better cycle efficiency and hence higher work output. On the other hand, the pinch exists at a temperature as high as possible in the context. The gas turbine therefore (Figure 2.71 (c)) has to reject a significant quantity of heat below the pinch, and so much of the work output is achieved at no more than the cycle stand-alone efficiency.

2.3.5.2. Best use of driving forces—Under the description of the Grand Composite Curve concept in section 2.2 of the Guide, it was shown how certain sections (shown shaded—see Figure 2.67 (a)) are "self balancing" in enthalpy terms. These sections clearly represent heat transfer at greater than minimum driving force, and on some processes, they can be large, with large included driving forces. In these situations it is sometimes worthwhile exploiting the opportunity for increasing power output at 100% efficiency.

Figure 2.72 (a) shows the source profile for a process which has a lot of spare heat available at high temperature (*i.e.* over ABC) but consumes rather a lot of heat in a low-temperature reboiler (DC). If we regard BCD as a self-contained area and just fit a Rankine cycle as shown, then large temperature differences remain untapped. If however, we remove DC from the Problem Table analysis, as shown in Figure 2.72 (b), effectively shifting DE to the right, then a lot more steam can be raised, with consequently much less waste of driving force. The load DC of course then remains to be satisfied, *which can be done using back-pressure steam from the Rankine cycle*. The net result is more of the "waste" heat below the pinch being converted into power, because the "spare" driving force existing between BC and DC in Figure 2.72 (a) is being exploited by the Rankine cycle. Care must be taken, however, not to exceed the load limit on the system. If DC is large enough, there may be insufficient back-pressure steam available, in which case heat from an external source would have to be brought in to make up the deficit. Other cases of this type are discussed by Townsend and Linnhoff (1982a).

2.3.6. Refrigeration systems

2.3.6.1. Comparison of basic types—A refrigerator is simply a heat pump, but one where the ultimate destination of the rejected heat is the ambient sink. Thus a simple "mechanical" refrigeration system is the same as the system shown in Figure 2.70, except that the working fluid is condensed against cooling water.

An alternative type of refrigerator is the absorption system, shown in Figure 2.73. Here, it is the "work potential" of above-ambient heat (usually steam) which effects the heat pumping without actually converting heat into shaft power. These systems tend not to be favoured nowadays due to their high capital costs (two columns required, one of them a high pressure column) and heavy heat demand.

However, the Grand Composite Curve, once again, can give the clue as to when the absorption system might be favoured over the ubiquitous compression system, as

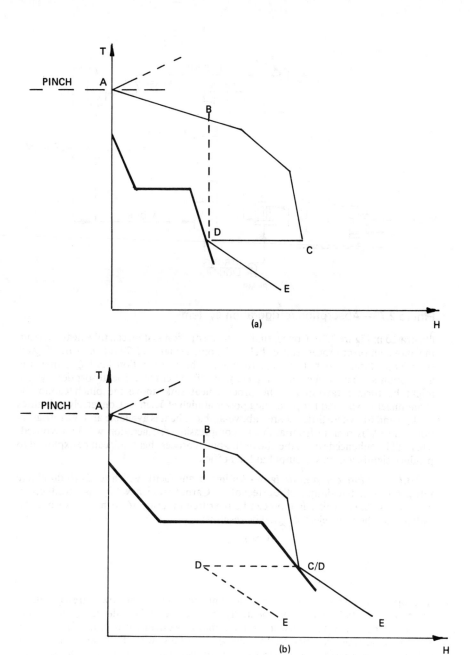

Figure 2.72—Best use of heat recovery network driving forces

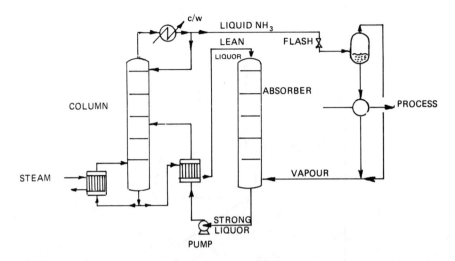

Figure 2.73—Absorption refrigeration system

illustrated in Figure 2.74. Conceptually, the absorption system can take heat Q_D from the above-ambient process source below the pinch (Figures 2.74 (a) and (b)). If Q_D is sufficiently large and hot enough to handle the refrigeration load Q_R, then the absorption system can be run completely on "free" energy. The absorption system might be further favoured if the process heat sink above the pinch cannot be economically exploited for producing power at high efficiency. Conversely in Figures 2.74 (c) and (d) where little "spare" above-ambient heat is available and the refrigeration load Q_R is remote from ambient, compression refrigeration will be favoured. This will be enhanced where the above-the-pinch process heat sink can be exploited to produce significant power output at high efficiency.

2.3.6.2. More complex designs—An important factor which tends to dominate refrigeration system design is the effect of the Carnot efficiency, "η_c" as introduced in equation 4 above. This equation can be rewritten in terms of the heat absorbed, Q_2 rather than the heat rejected, Q_1. The result is:

$$W \geqslant \eta_c . Q_2 \qquad \qquad \ldots\ldots(5)$$

where $\eta_c = \dfrac{T_1 - T_2}{T_2}$

The upper temperature, T_1 is fixed at ambient, but as the lower (refrigeration) temperature T_2 falls, η_c rises exponentially and becomes unbounded as T_2 approaches the absolute zero. Hence, refrigeration becomes very expensive in terms of power consumed as the required refrigeration temperature falls, with the power requirement being very sensitive to irreversibilities in the system design. For this reason, designers incorporate more complexity in low-temperature below-ambient heat and power systems than they do in above-ambient systems, in an effort to minimise power consumption.

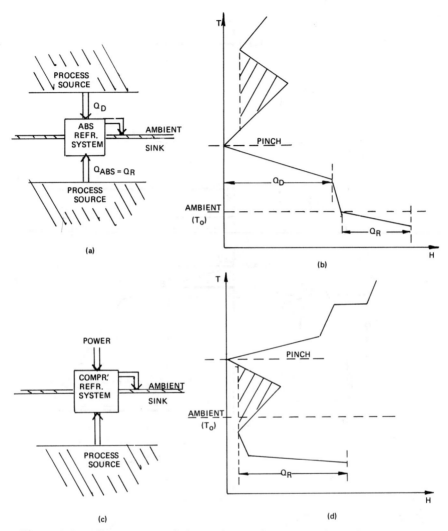

Figure 2.74—Comparison of absorption and compression refrigeration system applications

One way of reducing the power required by the simple device in Figure 2.70 is to incorporate an "economiser", as shown in Figure 2.75 (a). The compression and flash expansion are split into two stages, with flash vapour from the first expansion stage being returned to the suction of the highest pressure compressor stage. In this way, the quantity of vapour flowing through the lowest pressure part of the system is reduced, saving power.

Matching of refrigeration cycles against process source/sink profiles is illustrated in Figures 2.75 (a) to (f). Because one never cools above-ambient duties using refrigera-

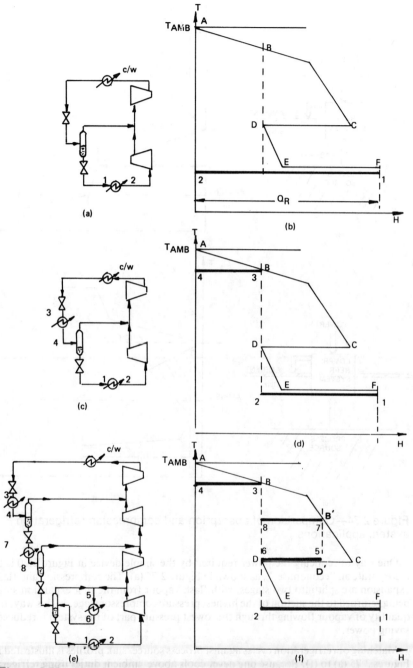

Figure 2.75—Matching compression refrigeration systems against the Grand Composite Curve

tion (for obvious reasons!), in any process, a pinch always exists (a "utility pinch") at cooling water temperature. This is point A in Figures 2.75 (b), (d) and (f). The cooling duty, below the cooling water pinch temperature must be handled by refrigeration. With the process source profile ABCDEF, the load Q_R could all be handled by the system in Figure 2.75 (a), with all process duties supplied from a single level 1−2 (Figure 2.75 (b)). However, with this design, large loss of driving force exists between AB and the refrigeration utility. Considerable power saving is achieved by the design in Figure 2.75 (c) where the process duty due to AB is moved to higher level 3−4, shown in the T/H diagram in Figure 2.75 (d). Shifting load upwards in temperature like this reduces vapour flow in the low pressure part of the refrigeration cycle, although increases it in the higher pressure part. The net effect, however, is reduced power consumption.

A further waste of driving force exists in the region BCD due to process interchange. This can be utilised as shown in Figures 2.75 (e) and (f). A level 5−6 is added to *recover* refrigeration from DD′, and a level 7−8 is added to replace the cooling previously supplied by DD′ to BB′ in process interchange. A further power saving is obtained, because now, part of the heat rejected by the process into the lowest level 1 to 2 can be disposed of at below-ambient temperature, *i.e.* over 5 to 6. This more than compensates for increased load 7−8 at higher level. Note the sharp increase in design complexity in going from the design in Figure 2.75 (c) to that in Figure 2.75 (e). Whether or not such a design evolution is worthwhile will depend entirely on the loads and levels involved. However, it is worth noting that designs of the type shown in Figure 2.75 (e) are commonly seen in low-temperature gas separation plants.

Figures 2.75 (a), (c) and (e) by no means tell the whole story in refrigeration system design. The reader can turn to, for example, Haywood (1980) for more information. The main point to note is that the below-ambient Grand Composite Curve can be used in the systematic exploration of the design options.

2.3.7. Total site heat and power systems
2.3.7.1. Problem scope—In the previous sections we analysed the interaction of power systems with process heat recovery networks *on a local basis*. Now we widen the view to take in site-wide combined heat and power systems, including cooling systems.

Figure 2.58 showed a schematic energy flow diagram of a chemical process site heat and power system. Considering this diagram, and understanding the "appropriate placement" concept described previously, the overall efficiency of the system can be seen as a function of two fundamental factors, namely,

(i)   the internal efficiency of the power-producing systems, and

(ii) the closeness of the match between the steam consumptions at the various levels and the sink profiles of the consumer processes.

We will consider both of these important factors here. However, another set of factors which do not relate directly to thermodynamics, but have a great impact on system economics, must be considered. These are the *constraints* on the system, which are,

(i)   the ratio of heat to power required by the site,

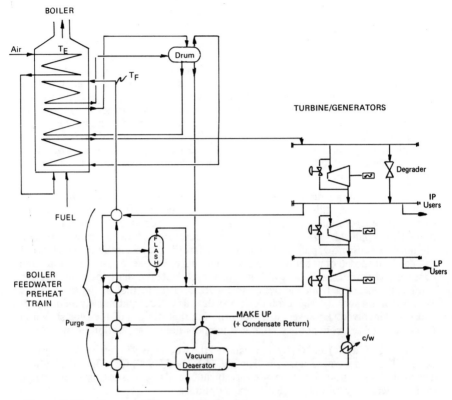

Figure 2.76—Power station steam Rankine cycle

(ii) the tariff structure imposed by the external electricity supplier,

(iii) the availability, type, and costs of fuels.

All of these have a fundamental impact on the strategy which should be adopted to improve (or design from scratch) the site heat and power system, and they will be discussed in the next section (2.4) of the Guide.

2.3.7.2. Steam power stations—As already stated the most common form of process power generation is the steam Rankine cycle. It can almost be guaranteed that this cycle will be the one chosen for the basis of any site central power station. The basic elements of the cycle are shown in Figure 2.76 (which can be compared with the simplified version presented in Figure 2.66). Steam is produced superheated and at high pressure by the boiler, and then expanded through a cascade of back-pressure power turbine stages. Site steam demand is met by steam passed out at IP and LP levels, and the lowest back-pressure steam level is used for deaerating incoming fresh makeup water and any returned condensate. A final expansion stage down to cooling water might be incorporated. Deaerated boiler feed water is pre-heated in a train of heat exchangers (utilising back-pressure steam and condensate, and heat from the boiler blowdown) up to the temperature T_F at which it enters the economiser in the boiler convection section. Because T_F will normally be quite high (perhaps around

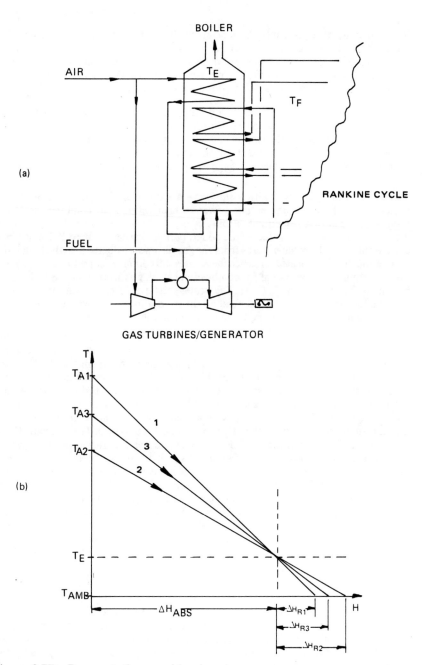

Figure 2.77—Power station combined cycle

200°C) then the flue gases in the boiler can only be cooled to a temperature somewhat higher than this by exchange against the feedwater. However, economic efficiencies for boilers today are in excess of 90%, which imply final flue gas exhaust temperatures, T_E, of around 140°C. Thus a fresh air pre-heater (as shown in Figure 2.76) must be included as the final bank in the stack if the economic value of T_E is to be achieved.

Many "variations on the theme" introduced in Figure 2.76 are used in practice, and the reader can refer to Haywood (1980) for more information. For example, additional back pressure levels can be added solely for the purpose of feedwater preheating (often known as "bled steam"). This increases the internal recirculation of steam in the system, thereby increasing power output and improving the cycle efficiency.

2.3.7.3. Combined cycle power stations—An important way of increasing power station cycle efficiency (increasing power to heat ratio) is to combine the steam Rankine cycle with an open gas turbine cycle, as shown in Figure 2.77 (a). Here, instead of burning all of the fuel in the boiler, part (which can be up to 100%) is burnt in a gas turbine. The exhaust gases from the gas turbine are ducted into the boiler and act as a source of pre-heated air, since these gases still contain around 14 vol.% oxygen. The air required by the system for combustion can all go through the gas turbine, or be split as shown in Figure 2.77 (a). The strategy to be adopted in deciding on how to split the fuel and air flows will depend on the ratio of heat to power required by the site (see later) and the economic trade-offs inherent in the system.

Figure 2.77 (b) illustrates some of the issues involved. It shows cooling curves in the T/H diagram for flue gas, for various cases of air flow split. Profile 1 is the case where all the air goes into the boiler (*i.e.* no gas turbine present). Assuming that fuel is burnt in the air to the maximum extent possible, then it cools from the adiabatic flame temperature T_{A1} down to ambient. Useful heat ΔH_{ABS} is absorbed by steam raising, down to a minimum allowed stack temperature T_E. Heat ΔH_{R1} is lost to ambient. If, however, all of the air is sent *via* the gas turbine, and fuel burnt in the turbine exhaust to the maximum extent possible, then flue gas profile 2 is obtained. The adiabatic flame temperature T_{A2}, is lower than T_{A1} because work has been extracted from the system by the gas turbine. Hence for a fixed steam-raising load and stack temperature, the air flowrate has to increase over the base case 1. This means that the quantity of heat lost to ambient ΔH_{R2} is increased, and the temperature driving force available for raising steam against the flue gas is reduced. Hence, case 1 represents minimum power output but lowest heat loss and lowest capital cost of steam-raising, and case 2 represents maximum power output but highest heat loss and highest capital cost of steam-raising. Profile 3 represents an intermediate case where the air flow is split. The economic choice depends heavily on the site constraints as listed under 2.3.7.1, above.

Note that, depending on the steam-raising pressure and flowrate, lowering driving forces as in profile 2 may actually mean that it is thermodynamically impossible to maintain ΔH_{ABS} fixed with T_E fixed. The flue gas flowrate may require to be raised still further, causing further increase in the heat loss to ambient. This situation will always be encountered when a gas turbine is added to an existing system, because of the fixed area in the boiler. In this situation a trade-off exists between modifying the boiler (adding area) and losing heat. The reader can refer to Cohen *et al* (1972) and Haywood (1980) for further details.

2.3.7.4. Distributed power generation—Perhaps one of the most important constraints on the site system, and the one which currently receives least attention, is the *topological* one of relying on a single power station for the supply of heat and electrical power. As stated earlier, this has the effect of forcing the heat consuming plants into accepting steam at the site main pressure levels. Thermodynamically these may not be the correct levels for an individual plant. Furthermore, the steam Rankine cycle (*i.e.* the site power station) might *not* be the most appropriate power cycle to match against a particular process. This was discussed with Figure 2.71 under 2.3.5.1, above.

This means that where a site contains processes which constitute sufficiently large heat sinks, *local* power-generating systems producing power for the site might well be able to supply better matched process heating than the central power station. This leads to increased power production at high efficiency, reducing the demand for power imported to the site from an external supplier. The techniques introduced in this section of the Guide should help the engineer identify, and quantify the benefits of, any opportunity of this type.

2.3.8. Summary

This section of the Guide has developed the important concept of Appropriate Placement of heat engines and heat pumps relative to the process pinch showing it to have a fundamental impact on the design of practical CHP systems. Further, it has shown that the process source/sink profile forms the basis of a design procedure for maximising power output of practical cycles at 100% efficiency. The following are important points to note in relation to this procedure.

(i)   The Grand Composite Curve forms a rigorous "thermodynamic reference point", against which practical power cycles can be matched by inspection. So although only steam Rankine cycles, gas turbine cycles and compression heat pumps, are discussed explicitly above, *any* type of cycle can be handled by analogous "profile matching" methods.

(ii)   As illustrated in Figure 2.71, different power cycles produce very different results when matched against the same process. Consequently the best practical cycle can be selected, with "experience" ΔT_{min} values assigned to the working fluid, *prior to detailed design*. Optimisation, in which the effect of varying ΔT_{min} on power output is studied, can be carried out at a later stage.

(iii)   Once the power cycle profile has been determined, it is added as a stream back into the process Problem Table analysis. A heat exchanger network incorporating the cycle working fluid can then be designed, as described in Section 2.2 of this Guide.

2.4. Economic Evaluation

Finally in this Section of the Guide, we address the question of economic evaluation of our "thermodynamically elegant" designs. Most of the comments are related to sites with a central heat and power utility system since it is in this context that economic assessments tend to be most tricky. First, we discuss the implications of constraints such as the site heat to power ratio (see Section 2.4.1), tariff structures (Section 2.4.2) and fuel values (Section 2.4.3). Then, we ask the question: Are there quick "back-of-the-envelope" type ways to assess savings or otherwise? The answer is: "Yes, there are", see Section 2.4.4.

2.4.1. The heat-to-power ratio

A fundamentally important constraint on many energy conservation schemes is the ratio of heat to power demanded by a site, and the levels at which the heat is demanded. These parameters determine which design is most economic. For example, with a high heat-to-power ratio, a steam Rankine cycle will probably give the best fit, whereas with lower ratios (*i.e.* more power required) the gas turbine/steam Rankine cycle combined system might be more appropriate. However, engineers are rarely in the position of designing whole site systems from scratch! More commonly, the engineer is faced with a system which was in balance when first built, but which has become out of balance due to subsequent changes in production patterns, and of course due to the energetic efforts of steam conservationists!

Consider the situation where the supply and demand of both process heat and power are roughly in balance, or where there is an excess of generating capacity as shown in Figure 2.78 (a). Any system changes which increase power output can only realise savings through sale of surplus power to the external utility supplier. Normally, rates obtainable are low, and sometimes the supplier simply will not accept exported power. Hence on sites where the process heat sinks dominate, one tends to see the excess steam demand being met by steam-raising "package" boilers. In this situation, savings in steam demands (*i.e.* those brought about by better heat exchanger network designs) simply save fuel in the package boilers. Substituting low grade steam demand for high grade cannot produce energy savings, neither can improving the internal power cycle efficiency. All that happens is that fuel burnt in the power station is replaced by fuel burnt in the package boiler!

In the opposite condition, where power demand exceeds the power generation capacity compatible with heat demanded by the process sinks, there is a definite incentive for improving power generation efficiency. If the power station is making up the power deficit by generating against cooling water in "condensing sets" as shown in Figure 2.78(b) rather than importing electricity (the economics of this are considered later) then cycle efficiency improvements can be used to "switch cooling water into fuel" (*i.e.* reduce total heat demand from fuel at constant power output). When electricity is being imported as in Figure 2.78(c), cycle efficiency improvements allow fuel to be "converted" to power at very high efficiency, (*i.e.* more power produced "in house" and less imported). When the power deficit is being made up by both "condensing sets" and power import as in Figure 2.78(d), cycle efficiency improvements allow "cooling water to be switched to power" (*i.e.* a combination of the cases in Figures 2.78(b) and (c)). However savings in the quantity of steam demanded by the site (through better process heat integration) realise reduced savings. This is because saving a tonne of steam not only saves the fuel required to raise it, but also eliminates the associated power output which is produced at around 87% efficiency (see under 2.4.4, below). Hence, since power generated at 87% must then be replaced by power generated at, say 30% (either on the site station or by the external supplier), the saving accruing from a tonne of utility steam saved is much less than that accruing from a tonne saved in a simple boiler.

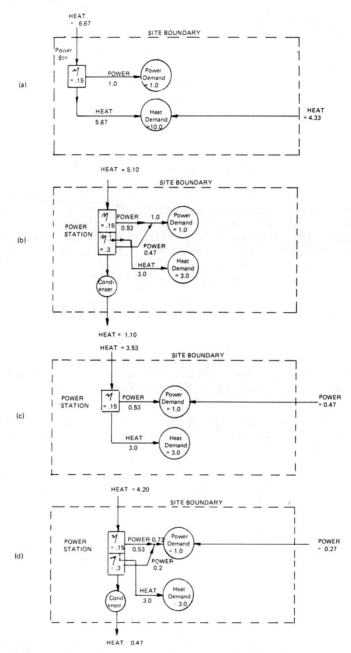

Figure 2.78—Heat and power supply and demand scenarios

119

2.4.2. Tariff structures

Another important constraint on many energy conservation schemes is the site tariff structure.

On sites dominated by power demand, the cost of imported electricity has a profound effect on the operating policy of the site power station.

Imported electricity does not normally have a single price. Rather, the price relates to the time of day when power is demanded. A typical (simplified) structure might be:

Night	(00-30 hrs to 07-30 hrs),	£11 per MW hr
"Shoulder"	07-30 hrs to 08-00 hrs and 00-00 hrs to 00-30 hrs,	£18 per MW hr
Day	(08-00 hrs to 24-00 hrs),	£25 per MW hr

Suppose this structure exists for a "power-deficit" site, which both imports power and has the ability to generate in-house power against cooling water, *i.e.* the situation in Figure 2.78(d). If the fuel available is fuel oil, costing £100 per tonne, and having a calorific value (net) of 39 900 kJ/kg, then the cost of heat from this source is:

$$\frac{100 \times 3600}{39\ 900} = £9.02 \text{ per MW hr}$$

So, if the "condensing" part of the power cycle has a cycle efficiency of 30%, then the marginal cost of in-house power is:

$$\frac{9.02}{0.3} = £30.07 \text{ MW hr}$$

This is much more expensive than imported power at any time of the day, hence it never pays to run the condensing sets. The power deficit should always be made up by importing.

We can also calculate what the marginal cost is of generating power against process sinks, rather than against cooling water. This is:

$$\frac{9.02}{0.87} = £10.37 \text{ MW hr}$$

where 87% is the marginal efficiency of power generation (see under 2.4.4 below).

In other words, it is only just worth doing during the 7 hour night period when imported electricity costs £11.00 per MW hr. A rise in the price of oil of only 10% would mean that at night it would not be worth producing *any* in-house power! The power station would simply be used as a source of process heat, with the turbines just "ticking over".

Finally it must be noted that some power contracts include a (power) "load-shedding" agreement. At certain times of day when the load on the external utility supplier is high, the site may have to reduce its electricity demand at short notice to an agreed minimum. Any electricity imported over and above the minimum during a load-shedding period incurs a severe cost penalty (£17 000 per MW hr is a typical figure!). It might well not be feasible in this situation for the site to shed enough power demand at the short notice available, in which case it pays for the power station to generate not only in the condensing sets, but also by blowing off back pressure steam to atmosphere!

2.4.3. *Fuel value*

Next, we must never lose sight of the fact that a unit of heat (*i.e.* 1 MJ) can have quite widely differing costs depending on the fuel source used.

Normally, a site will have more than one fuel available, with (on a common energy basis) different prices. Over recent years in the UK, fuel gas has had a considerable cost advantage over fuel oil. Hence the trend has been to use gas as a "base load" fuel (burnt to the maximum quantity allowed by the gas contract) and fuel oil as the "marginal" fuel for any excess heating duty. If, however, the whole of a site demand can be met by gas, *i.e.* gas becomes the marginal fuel, then it becomes economically attractive to generate power in-house using condensing sets during the high daytime power tariff period (see previous subsection). However, government policy in the UK is to discourage this by raising the gas price, thus creating an incentive for industry to switch to coal as the base load fuel. Currently, coal costs around £7.00 per MW hr of heat, making it attractive to use coal for in-house power generation in condensing sets during the day if the coal can be made the marginal fuel.

2.4.4. *"Back-of-the-envelope" assessments*

Finally, how can we quickly and confidently establish whether or not a "heat exchanger network" or "heat and power" design will save money rather than energy, given all the constraints discussed above and perhaps others? There are two concepts worth highlighting as convenient vehicles for fast evaluation: that of efficiencies and that of "drawing boundaries". Their use is now demonstrated by way of examples.

There are many ways of "tinkering" with complex systems and it is easy for the designer to get lost in circular arguments when trying to assess the effect of changes. It is therefore very important to keep in mind the simple concept of "crossing the boundary", as illustrated in Figures 2.79(a) and (b). The system boundary in Figure 2.79(a) encloses any notional heat engine (*i.e.* the same simple representation as used in the previous section of the Guide). Suppose we find a design change that increases the cycle efficiency. Then by consideration of the three energy flows which cross the system boundary, we have (by energy balance) exactly three ways that the improvement can be exploited:

(i) Keeping power output constant, the fuel input is reduced and hence the exhaust heat is reduced by the same amount, *i.e.* exhaust heat is converted to fuel at 1:1.

(ii) Keeping exhaust heat constant, extra power is produced from fuel at 1:1 (*i.e.* the effect obtained in the "Appropriate Placement" concept).

(iii) Keeping the fuel input constant, more power is obtained at the expense of exhaust heat, *i.e.* exhaust heat is converted to power at 1:1.

Which of these strategies is economically the best depends on system constraints, as discussed previously.

The above arguments relate to the main energy flows of the power cycle. However, in real systems, other small flows exist, which we will describe as system losses. These

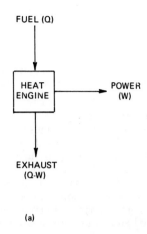

(a)

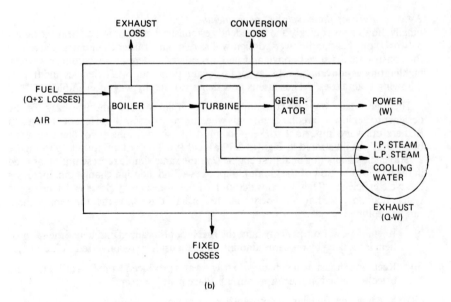

(b)

Figure 2.79—"Cross the boundary" analysis of CHP systems

are shown in Figure 2.79(b) for the steam power system, and can be classified into fixed and variable losses. Fixed losses are losses from hot surfaces, steam leaks from turbine glands, *etc*.

Losses which depend on power output are classed as variable losses, the chief of which are the stack loss in the boiler and the loss associated with mechanical transmission and power conversion. Although these losses do not scale in simple proportion with power output, the assumption that they do is normally good enough for economic potential calculations. Hence we can assign conversion efficiencies to the boiler and power conversion, typical values being 90% and 97% respectively. This means that when this type of system is "appropriately placed" relative to a process heating duty, then power is produced, not quite at 100% efficiency (case (ii) above), but at:

0.9 × 0.97 × 100 = 87% efficiency from the primary fuel.

The other two cases given above can be analysed similarly, *i.e.*

(i)   Power output constant but fuel reduced. Hence, conversion efficiency of exhaust heat into fuel is at:

$$\frac{100}{0.9} = 111\%$$

(iii) Fuel input constant, but power increased. Hence, conversion efficiency of exhaust heat into power is at:

$$0.97 \times 100 = 97\%$$

(provided the boiler air flow is not significantly altered).

It is clear from the previous three subsections that the economic potential of an energy conservation scheme will depend on the operating policy of the power station, and on the marginal fuel employed. Essentially, this means that the effect of schemes *on the energy flows across the site boundary* should be evaluated, and the resulting marginal costs calculated. This contrasts with the way that energy schemes are often evaluated in practice, by using the site internal transfer pricing structure. This latter procedure is only acceptable if the internal steam and power prices accurately reflect the site operating policy. However, transfer pricing is a notoriously "thorny" area, and it is a good idea to check any schemes evaluated, by using the "cross the boundary" method.

As another example, we now evaluate the effect of adding a gas turbine to a steam cycle power station, in the manner shown in Figure 2.77(a). The gas turbine has the following characteristics:

Power output = 20 MW

Overall cycle efficiency = 30%

Combustion chamber efficiency = 98%

Power conversion efficiency = 97%

Suppose that when this gas turbine is integrated with the power station boilers, the marginal boiler efficiency becomes 87%. If the use of the gas turbine saves imported power, then the marginal power generation efficiency is:

$$0.98 \times 0.97 \times 0.87 \times 100 = 82.7\%$$

Hence the marginal fuel required is:

$$\frac{20}{0.827} = 24.18 \ MW$$

The gas turbine fuel demand is:

$$\frac{20}{0.3} = 66.67 \ MW$$

The fuel burnt in a gas turbine is preferably a lighter, "high grade" fuel. Fuel gas is ideal. Hence there must be the equivalent of 66.67 MW of fuel gas continuously available for firing on the gas turbine. Assuming that fuel oil is the marginal fuel, we can now calculate the economic potential of the scheme. Using the tariff structure introduced previously, with fuel oil costing £9.02 per MW hr, then marginal savings against imported electricity are:

Night: $11.00 - \dfrac{9.02}{0.827}$ = £0.093/MW hr
                                            for 2555 hrs per year

Shoulder: $18.00 - \dfrac{9.02}{0.827}$ = £7.093/MW hr
                                            for 365 hrs per year

Day: $25.00 - \dfrac{9.02}{0.827}$ = £14.093/MW hr
                                            for 5840 hrs per year

Hence, for 20 MW generated, total savings are:

$$20 \,(0.093 \times 2555 + 7.093 \times 365 + 14.093 \times 5840) = \text{£}1.7 \times 10^6 \text{ per year}$$

An aero-engine derivative gas turbine/alternator set of this size would cost about £1.8 × 10⁶ at current (1982) prices, typically requiring an installation factor of about two. Hence the total installed capital cost would be about £3.6 to £4.0 million, yielding a simple payback of two to three years. This result is typical of current "gas turbine schemes".

List of Symbols*

A	Heat transfer area (m²)
CP	Heat capacity—flowrate (kW/K)
F_i	Heat flow through heat recovery network at level i
H	Flow enthalpy (kW)
ΔH	Change in flow enthalpy (kW)
L	Number of loops in a network (dimensionless)
N	Number of process streams plus utilities, or process stream branches (dimensionless)
Q	Heat flow (kW)
$Q_{H\min}$	Minimum feasible hot utility (kW)
$Q_{C\min}$	Minimum feasible cold utility (kW)
R	Heat rejected below the pinch by heat engine (kW)

*N.B. Most "case studies" and examples in the text use "engineering" units (*i.e.* °C, °F, te cal/hr, BtU/hr)

S	Entropy (kJ/kgK)
s	Number of separate components in a network (dimensionless)
T	Temperature (K)
T_S	Supply temperature of process stream (K)
T_T	Target temperature of process stream (K)
ΔT	Temperature difference (K)
ΔT_{min}	Minimum allowed temperature difference (K)
$\Delta T_{threshold}$	Largest value of ΔT_{min} that can be used in threshold problem design (K)
$\Delta T_{L.M.}$	Log mean temperature difference (K)
U	Heat transfer coefficient (kW/m²K)
u	Number of heat exchange units (*i.e.* heaters, coolers, exchangers) (dimensionless)
u_{min}	Minimum number of units (dimensionless)
$u_{min\ MER}$	Minimum number of units for maximum energy recovery (dimensionless)
X	Heat load shifted around a loop or along a path (kW)
α	Heat flow across the pinch (kW)
η	Heat engine efficiency (dimensionless)
η_c	Reversible heat engine efficiency (dimensionless)
ϕ	Heat pump coefficient of performance (dimensionless)

Subscripts

A	Adiabatic
AMB	Ambient
ABS	Absorbed heat
H, HOT	Relating to hot stream
C, COLD	Relating to cold stream
D	Driving heat
E	Exhaust
F	Feed condition
R	Rejected heat

1, 2, $\cdots$ A, B $\cdots$ i, n, counters

Glossary of Terms

Appropriate Placement: Matching of heat engines or heat pumps relative to the heat recovery pinch for best combined energy performance.

Cascade: Set of heat flows through a heat recovery problem, in strict descending temperature order (as calculated in Problem Table analysis—see **Problem Table**).

Cold Stream: Process stream requiring heating.

Composite Curve: Combined temperature—enthalpy plot of all hot or cold streams in a problem.

CP-**Table:**   Tabulated values of stream heat capacity flowrates, immediately above or below the pinch.

Cyclic Matching: Repeated matching of pairs of process streams.

Data Extraction: Definition of data for energy integration studies, from a given flowsheet.

Energy Relaxation: Process of reducing energy recovery in a heat exchanger network for the purpose of design simplification.

Grand Composite Curve: Plot of heat flow *versus* temperature from a heat cascade (see **Cascade** and **Problem Table**).

Grid: System of horizontal and vertical lines with nodes, for representing heat exchanger networks.

Heat Exchanger Network: System of utility heaters and coolers and process interchangers.

Hot Stream: Process stream requiring cooling.

Loop: System of connections in a heat exchanger network which form a closed pathway.

Maximum Energy Recovery (MER):   Best possible energy recovery in a heat exchanger network for a given value of ΔT_{min}.

Multiple Utilities: Utility or utility system whose temperature or temperature range falls within the temperature range of the process stream data.

Path: System of connections in a heat exchanger network forming a continuous pathway between the utility heater and a utility cooler.

Pinch: Point of zero heat flow in a cascade (alternatively, point of closest approach of composite curves in a "heating and cooling" problem).

Pinch Design Method: Method of heat exchanger network design which exploits the constraints inherent at the pinch.

Pinch Match: Process interchanger which *brings* a stream to its pinch temperature (*i.e.* hot streams above the pinch, cold streams below).

Problem Table: System of analysing process stream data for a heat recovery problem which exploits **Temperature Interval** sectioning of the problem, and predicts minimum utilities consumptions, pinch location, and cascade heat flows.

Process Sink Profile: Section of the Grand Composite Curve above pinch temperature.

Process Source Profile: Section of the Grand Composite Curve below pinch temperature.

Profile: Temperature-enthalpy plot of a stream or a composite stream.

Revamp: Any change to an existing chemical process, but in this context, mostly changes for improvement in energy efficiency.

Stream-Splitting: Division of a process stream into two or more parallel branches.

Subset: Set of process streams or process streams, plus utilities, within a heat recovery problem which are in overall enthalpy balance.

Supply Temperature: Temperature at which a process stream enters a heat recovery problem.

Target: A design performance limit, determined prior to design.

Target Temperature: Temperature at which a process stream leaves a heat recovery problem.

Temperature Interval: Section of a heat recovery problem between two temperatures which contains a fixed stream population.

Tick-Off-Rule: Heuristic of maximising the heat load on an interchanger.

Threshold Problem: Heat recovery problem that shows the characteristic of requiring either only hot or only cold utility, over a range of ΔT_{min} values from zero up to a threshold.

U.A. **Analysis:**   Procedure of calculating UA values ($= Q/\Delta T_{LM}$) for matches in a heat exchanger network, for the purposes of preliminary costing and optimisation.

Utility: System of process heating or process cooling.

Unit: Process interchanger, heater or cooler.

ΔT_{min} **Contribution:**   Temperature difference value assigned to individual process streams. Match-dependent ΔT_{min} values are given by the sum of the contributions in a match.

3. Heat Transfer Equipment

3.1. Introduction

3.2. Initial Selection

3.3. Selection between Feasible Types

3.4. Performance and Cost Information on the Various Heat Exchanger Types

3.4.1.   Shell-and-tube heat exchangers

3.4.2.   Double pipe heat exchangers

3.4.3.   Gasketed plate heat exchangers

3.4.4.   Spiral heat exchangers

3.4.5.   Lamella heat exchangers

3.4.6.   Welded plate exchangers

3.4.7.   Hot gas-to-liquid convective bank systems

3.4.8.   Gas-to-gas recuperative heat exchangers

3.4.9.   Heat-pipe heat exchangers

3.4.10.   Rotary regenerators

3.5. Localised Utilities

3.5.1.   Air-cooled heat exchangers

3.5.2.   Cooling towers

3.5.3.   Package boilers

3.6. Examples of Application

3.6.1.   Oil cooler

3.6.2. Process water cooler

3.6.3. Light hydrocarbon condenser

3.6.4. Comparison of air cooled heat exchanger and cooling tower

3.1. Introduction

As will have been seen from the previous sections, heat exchangers are key equipment items for process integration.[1] The objectives of Section 3 are as follows:

(1) To present information on heat exchangers enabling the process designer to select the appropriate *type* of heat exchanger for his particular application. The first stage in the selection procedure is to identify the exchanger types which might be suitable for the application in terms of operating parameters such as temperature, pressure, corrosive nature of the fluids *etc*. This initial selection will normally identify several types from which a final selection can be made, usually on a cost basis. This section concentrates on those cases where a potential choice exists and does not deal with situations where the process requirements dictate absolutely the type of unit which must be used; an example in this latter category is that of the scraped surface heat exchanger which is usually only employed when the conditions (*e.g.* in crystallisation) are such that no other system is feasible.

(2) To provide approximate cost data (in a new form) on the various heat exchanger types. This cost data is important not only in selecting the type from the list of physically feasible types, but also in determining whether a *prima facie* case exists for the application of heat transfer in the particular energy conservation application and for carrying out a preliminary optimisation.

In pursuance of the above objectives, this section starts by presenting the range of possible types in tabular form indicating the limiting parameters such as pressure, temperature *etc*. Next, the procedure for selection between feasible types is outlined, with definitions begin given of heat transfer rate, temperature difference, overall heat transfer coefficient and a new parameter C which is the cost per unit heat transfer per unit temperature difference ($£/(W/K)$). The next part of the section (3.4) presents information on each of the types mentioned in the initial selection table, giving values of overall heat transfer coefficients and C values. The final part of the section gives several examples of application of the proposed methods.

It should be stressed at the outset that the methods given in this section are approximate in nature. They are useful in giving an overall "engineering feel" for the design, but their accuracy is strictly limited and where several alternative designs give around the same cost, it is certainly worthwhile to seek quotations for each of the alternatives. Other factors which are important in design such as pressure drop and vibration are not specifically accounted for. The coefficients and C values stated are those appro-

---

1 Therefore, it is important to identify quickly an appropriate type of exchanger and to establish its approximate cost. Clearly, this step can be seen as a vital part of the iterative approach to process integration.

priate to the normal range of design for the process industries. Where a particularly high pressure drop is available across the heat exchanger or where the pressure drop must be reduced to an unusually low value, then the costs can be significantly lower or higher than those indicated.

Smith (1981) shows that the total cost of heat exchangers (annual pumping cost plus annualised capital cost) is a function of fluid velocity. A minimum total cost occurs at an optimum fluid velocity combination. However, the optima identified by Smith are rather flat and this fact makes the much simpler approach adopted here more useful than might have been supposed. Except for the on-site assembly costs of some units, installation and pumping costs have not been included in the costs stated in this chapter. Such costs will vary from situation to situation and should be borne in mind in making comparisons.

It cannot be stressed too strongly that the methods given in this section cannot possibly provide general solutions. There is no substitute for detailed design taking account of all the relevant factors. Nevertheless, it is hoped that, by presenting information and approximate cost data on a variety of heat exchanger types, it will be possible to persuade process designers to take a broader view of the application and selection of heat exchangers.

The provision of utility (heating system, cooling water) was discussed in Section 2. At a given site, the cost of utilities is often specified on a tariff basis. An alternative is to use *localised* utilities (package boilers, air-cooled heat exchangers, small cooling towers) and the choice between localised and site supply will obviously depend on relative costs. Section 3.5 gives a brief review of localised utility plant.

3.2. Initial Selection

The range of heat exchangers considered in this present section is listed in Table 3.1. The list is not completely exhaustive, but it does include the more commonly used types. For the particular application being considered, the list given in the table should be checked for the following:

(1) *Maximum pressure*—many exchanger types operate only at low pressure and can immediately be ruled out for particular applications.

(2) *Temperature range*—many exchanger types are applicable over only a limited range of temperature and, again, this rules out a number of types.

(3) *Fluid limitations*—here the main emphasis is on compatibility between the fluid and the materials of construction. For instance, in gasketed plate exchangers, it may not be possible to find a gasket material which will be compatible with the fluid. Another factor is that concerned with the consequences of failure leading to interstream mixing and/or leakage. Clearly, for highly toxic or inflammable fluids, the consequences of failure are much more significant and this can often be an important consideration in deciding on the heat exchanger type.

(4) *Size range available*—obviously, it is always possible to overcome the problem of maximum size limitation by having a number of heat exchangers in parallel. However, this itself gives additional costs in piping and sometimes leads to problems in the distribution of flow between the parallel units.

Table 3.1

Exchanger type	Maximum pressure	Temperature range	Fluid limitation	Normal size changes for individual unit	Special features
Shell and tube	30.7 MPa (for higher pressure codes not applicable)	− 200 to ~ 600°C (higher with special materials)	Subject only to materials of construction	10 to 1000 m² (per shell—multiple shells can be used)	Very adaptable and flexible can be used for nearly all applications
Double pipe (plain and finned tubes)	> 30.7 MPa (shell) > 140 MPa (tube)	− 100 to ~ 600°C	Subject only to materials of construction	0.25 to 200 m²	Standard modular construction
Gasketed plate exchangers	1.6 MPa (possible to group to 2.5 MPa depending of size and temperature)	− 25 to 175°C (− 40 to 200°C possible for very special types)	Normally unsuitable for gases two phase flow. Limitation is on gaskets	1 to 1200 m²	Modular construction. Normally the most economical if applicable
Spiral heat exchangers	1.8 MPa	Up to 400°C	Subject only to materials of construction. May be suitable for fouling duties	Up to 200 m²	Easily tailored to process specification. High heat transfer efficiency. Low maintenance—self cleaning. Low installation cost
Lamella heat exchangers	2.0 MPa	Up to 220°C (Teflon gasket). Up to 500°C (asbestos gaskets and stainless steel)	Particularly suitable for gas-to-gas duties. Subject only to materials of construction	1 to 1000 m²	High thermal efficiency (counter current flow). High coefficients straight flow paths minimum solid deposition

Type	Pressure	Temperature	Application	Surface area	Comments
Welded plate exchangers	3.0 MPa (higher in shells)	In excess of 400°C	Subject only to materials of construction. Not suitable for fouling duties	> 1000 m²	Differential pressure should be less than 3.0 MPa
Hot gas-to-liquid convection bank systems	Near atmospheric (gas side). High pressure possible on process side	Up to 700°C	Usually used for water heating (economisers). Also used for other liquids	Up to around 500 m²	Widely spaced rectangular fins to minimise fouling or plain tubes
Gas-to-gas exchangers	Near atmospheric (shell side). Higher on tube side	Depends on type. Typically 250°C, others higher	Gases. Typically waste heat streams	Lower temperature forms typically 6 to 100 m² cast iron element recuperators 1200 to 3000 m²	Wide variety of types available selection depends strongly on corrosive nature of gas
Heat pipe and thermosyphon heat exchangers	Near atmospheric	~200°C (higher with special heat pipe fluids)	Low pressure gases	100 to 1000 m²	Near counter-flow operation possible. Extended surface possible on both sides
Rotary regenerators	Near atmospheric	Up to 980°C	Low pressure gases		Inter-stream leakage must be tolerated
Air cooled heat exchangers	High on process side	High temperature possible on process side	Subject only to material of construction	5 to 200 m²	Heat rejection system. Highly standardised design

In addition, the question should be asked, "Is there a temperature cross?" (*i.e.*, is the outlet temperature of the hot fluid higher than the inlet temperature of cold fluid?) If so, units approximating counter-current flow are—needed either pure counter-flow or multi-pass units in series; if not, cross-flow or mixed flow units are adequate.

Examination of Table 3.1 should lead to identification of at least one, and probably several, types of exchanger which are suitable for the particular application.

3.3. Selection Between Feasible Types

Having produced from Table 3.1 a list of feasible heat exchanger types for the particular application, the next step is to obtain an indication of the cost of heat exchanger for each of these types.

The rate of heat transfer Q (Watts) in a heat exchanger may be calculated from the expression:

$$Q = W_1 (h_{1i} - h_{1o}) = W_2 (h_{2o} - h_{2i}) \qquad \dots \dots (1)$$

where W_1 and W_2 are the mass rates of flow (kg/s) of the two fluids, with heat being transferred from fluid 1 (the hot side fluid) to fluid 2 (the cold side fluid). h_{1i} and h_{2i} are the inlet specific enthalpies of the two streams and h_{1o} and h_{2o} are the outlet specific enthalpies of the streams. For fluids of constant specific heat, Q may be derived from the expression:

$$Q = CP_1 (T_{1i} - T_{1o}) = CP_2 (T_{2o} - T_{2i}) \qquad \dots \dots (2)$$

where CP_1 and CP_2 are the heat capacity flowrates of the respective fluids. Here, $CP_1 = W_1 c_{p1}$ and $CP_2 = W_2 c_{p2}$ where c_{p1} and c_{p2} are the specific heat capacities of the respective fluids. T_{1i} and T_{2i} are the inlet temperatures and T_{1o} and T_{2o} are the respective outlet temperatures.

Conventionally, we may write:

$$Q = UA \Delta T \qquad \dots \dots (3)$$

where A is the surface area of the heat exchanger (normally taken as the surface in contact with one or other streams; for example in shell and tube heat exchangers the convention is to take the surface area as that corresponding to the outside of the tube) ΔT is an effective temperature difference between the streams and the constant of proportionality U is referred to as the *overall heat transfer coefficient*. The most efficient form of heat exchanger is one which operates in pure counter-current flow as illustrated in Figure 3.1. For this case, and with the assumptions of constant fluid

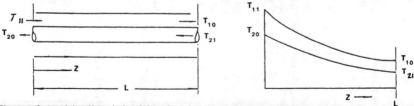

Figure 3.1—Idealised double-pipe heat exchanger with counter-current flow

134

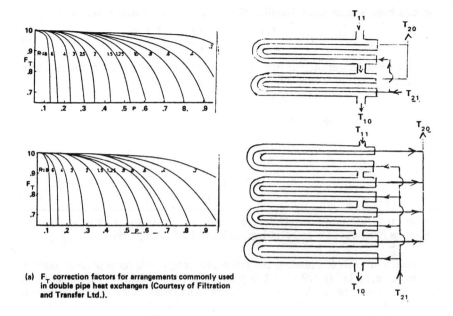

(a) F_T correction factors for arrangements commonly used in double pipe heat exchangers (Courtesy of Filtration and Transfer Ltd.).

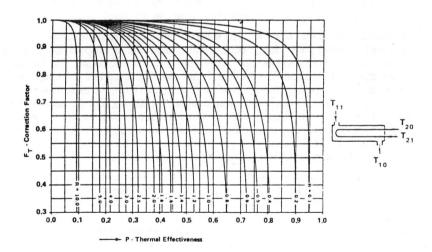

(b) F_T correction factors for shell - and - tube heat exchangers with an even number of tube side passes and with the shell side fluid uniformity mixed in a given cross section.

Figure 3.2—F_T correction factors for tubular exchangers

specific heats and constant overall coefficient along the exchanger, we have:

$$\Delta T = \Delta T_{LM} = \frac{(T_{1i} - T_{2o}) - (T_{1o} - T_{2i})}{\ln[(T_{1i} - T_{2o})/(T_{1o} - T_{2i})]}$$ . . . . . (4)

Many heat exchangers depart from the purely counter-current flow form and, here, we define the effective temperature difference as follows:

$$\Delta T = F_T \Delta T_{LM}$$ . . . . . (5)

where F_T is a correction factor which is often expressed in terms of the ratios:

$$R = \frac{T_{1i} - T_{1o}}{T_{2o} - T_{2i}}$$ . . . . . (6)

$$P = \frac{T_{2o} - T_{2i}}{T_{1i} - T_{2i}}$$ . . . . . (7)

Values of F_T for shell-and-tube type units and for cross flow (without mixing) are shown as functions of R and P in Figures 3.2 and 3.3 respectively.

A much fuller set of equations and graphs, covering a wide range of cases, is given, for instance, by TEMA (1976) or Taborek (1982). The sensitivity of F_T to small variations in P in some ranges should be noted and is an important factor in deciding between alternative designs. It is important to note the considerable adverse effect that back mixing of either the hot or the cold fluids can have upon the value of F_T.

The value of F_T may be increased for shell-and-tube exchangers by having shells in the series. However, this may increase the overall cost.

It should be noted that U is not necessarily constant along the heat exchanger; this is particularly so when phase change occurs and condensation or evaporation gives rise to a variation of velocity and local parameters along the heat exchanger. Furthermore, the assumption of constant specific heat is not always obeyed and the temperature profiles along the heat exchanger can be quite complex. Thus, the basic assumptions underlying the use of a logarithmic mean temperature driving force (equation 4) are often not obeyed. However, at the level of sophistication being used in this present section, the use of corrected logarithmic mean temperature differences is probably of sufficient accuracy. Of course, these effects must be taken account of in the accurate design necessary to reach a firm quotation for a particular duty.

The overall heat transfer coefficient U will vary with fluid enthalpy (particularly in multiphase systems) and with fluid velocity. There is a fundamental objection, therefore, to specifying a given value for a particular duty and configuration. However, long experience has shown that for designs around the normal value, order-of-magnitude estimates can be made for U and that these are extremely useful in doing quick calculations of heat exchanger size and for checking for gross errors in calculations. Often, in the heat transfer text books, ranges of typical values are given. In the present section, we have chosen to give a single value as being typical of a particular duty and configuration. There is no implication that the assigned value is an accurate one, it is given merely for the purpose of quick order of magnitude assessment. The use of this procedure has a time-honoured place as a precursor to detailed design.

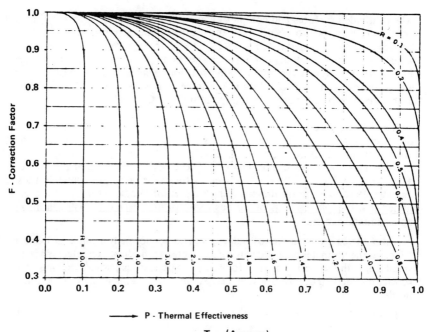

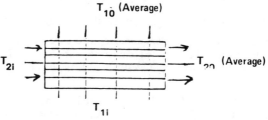

T_{10} (Average)

T_{2i} T_{20} (Average)

T_{1i}

Figure 3.3—F_T correction factor for an exchanger for pure cross flow with no lateral mixing in the cross flow stream

Where a very large range of conditions is covered in one unit, the unit can be considered as a series of connected units, each with its own representative conditions.

Consideration needs to be given to the in-service value of U after fouling has taken place, allowing for the mitigating effects of using built-in means of cleaning (e.g. soot blowers).

If U can be estimated approximately, then the heat exchanger area A may be calculated from:

$$A = \frac{1}{U}\left(\frac{Q}{\Delta T}\right) \qquad \ldots\ldots(8)$$

The quotient $(Q/\Delta T)$ is characteristic of the heat exchanger duty being carried out and the cost of the exchanger to perform the duty is often estimated by multiplying A by a cost per unit area. A difficulty arises here over the definition of area, particularly when extended surfaces and complex geometries are employed. However, from the point of view of the process designer, the really key question is the overall cost for his particular duty, specified in terms of $(Q/\Delta T)$. Here, we have defined a cost factor C which represents the cost in pounds sterling per unit $(Q/\Delta T)$; C has the units £/(W/K). For a particular duty and configuration, therefore, values of C may be estimated and are given in addition to U values in the table in this section. The cost of the heat exchanger may be estimated by simply multiplying C by $(Q/\Delta T)$.

Normally, the cost of a heat exchanger per unit surface area, and hence the C value, decreases with increasing heat exchanger size. In the information presented here, C is given at specified values of $Q/\Delta T$ and its value at intermediate levels of $Q/\Delta T$ may be estimated by logarithmic interpolation. The costs of heat exchangers are clearly affected by inflation. The figures given in the tables are approximately those for January 1982, but clearly these will increase as inflation continues. However, it is hoped that the relative costs of various forms of heat exchanger will continue to be approximately correct.

It should be noted that the cost values given do not normally include site installation. Whereas the normal piping and valving associated with a heat exchanger is a relatively small item, the same is not true for hot gas applications, where ducting, perhaps refractory lined, and, possibly dampers, capable of working at high temperatures are needed.

To summarise, the procedure for evaluation of the alternative feasible types of heat exchanger identified in the initial selected procedure, is as follows:

(1) The heat load Q should be estimated using equations 1 or 2.

(2) The logarithmic mean temperature difference ΔT_{LM} should be estimated from equation 4 and the corrected temperature difference ΔT estimated by multiplying ΔT_{LM} by F_T, which may be estimated from Figures 3.2 and 3.3 and equations available in the literature.

(3) The quotient $Q/\Delta T$ should be calculated for each proposed configuration (note that for the particular duty required, the values of $Q/\Delta T$ will vary since the temperature difference correction factor F_T varies).

(4) From the tables provided below for each exchanger type, read off the value of C, interpolating logarithmically between the levels of $Q/\Delta T$ given in the tables.*

(5) Calculate the cost of each configuration for the specified duty by multiplying $Q/\Delta T$ by C and compare the costs, bearing in mind possible differences in installation and pumping costs.

(6) If one configuration is greatly better than the others (a factor of 1.5, say) then this design should be selected and detailed calculation and estimates carried out. If there are several designs at around the same cost, then all of the designs should be estimated in greater detail.

---

* A convenient graph for this interpolation is given as Figure 3.24 in the context of the worked examples.

3.4. Performance and Cost Information on the Various Heat Exchanger Types

3.4.1. *Shell-and-tube heat exchangers*

The shell-and-tube heat exchanger is the most commonly used type in the process industry accounting for at least 60% of all heat exchangers used. The reason for its ubiquitous application is that it can be designed to operate over the full range of pressures and temperatures encountered in the process industry, it can be constructed from a wide range of materials, there exist well established design methods and mechanical codes for the units and there have been many decades of experience in their application leading to a high degree of confidence.

The most commonly used types of shell-and-tube heat exchanger are illustrated in the inset in Table 3.2 and are as follows.

(1) *Fixed tube plate*—This type of exchanger is probably used more often than any other type. The construction is simple and economical and the tube bores can be cleaned mechanically or chemically. However, the outside surface of the tubes are inaccessible except to chemical cleaning. If large temperature differences exist between the shell and tube materials, it may be necessary to incorporate an expansion bellows in the shell, to eliminate excessive stresses caused by expansion.

(2) *U-tube type*—Heat exchangers of this design are usually more expensive than the fixed tube sheet design, but the differences are often small and U-tube units can be cheaper at high pressure. The tube bundle can be removed from the shell for cleaning or maintenance. Internal cleaning of the tubes by mechanical means is difficult and it is usual to apply this type where the tube side fluids are clean. Thermal expansion is absorbed by the U-bends. The number of tube sheet joints is reduced, but tube replacement is more difficult than with straight tube exchangers since it may be necessary to remove a number of tubes before a tube in the centre of the bundle can be replaced. A disadvantage of the U-tube design is that it cannot have pure counter-flow (except with F-type shells) and this gives a consequential penalty on effective ΔT.

(3) *Floating head type*—This type of heat exchanger is suitable for the rigorous duties associated with high temperatures and pressures. The tube bundle consists of straight tubes secured into tube plates at each end. The floating head is fitted to the tube plate remote from the channel end and the cover is held in position by a split backing ring. The floating head tube plate is only slightly smaller in diameter than the inside of the shell and the floating head can only be assembled after the tube bundle is in position in the shell. Conversely, the floating head must be dismantled before bundle removal. The tube bundle is easily removed for cleaning or maintenance and the tubes can be mechanically cleaned both inside and outside. The floating head design is more expensive (approximately 25%) than the fixed tube sheet design.

The types shown in Table 3.2 are the three most commonly used, but many other configurations are possible and are used from time to time. The combinations are usually described in terms of a standard nomenclature devised by the Tubular Exchanger Manufacturers Association (TEMA) of the USA. The TEMA nomenclature is defined in Figure 3.4. It will be seen that the three types shown in Table 3.2 are (in terms of the TEMA nomenclature) designated BEM, BEU and BES respectively. The first letter describes the front end type, the second letter the shell type and the third letter the rear end type.

Shell and Tube Heat Exchanger

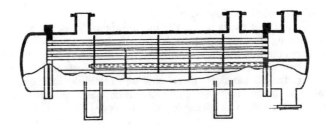

FIXED TUBEPLATE (TYPE BEM)

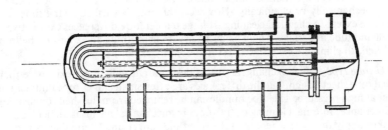

U-TUBE (TYPE BEU)

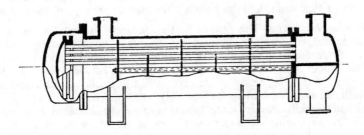

FLOATING HEAD (TYPE BES)

Note: #   Unit needs multiple shells to meet area.

          *   Viscosity range 1 – 5 centipoise
          *   Viscosity range > 100 centipoise
          ***   Viscosity typically < 1 centipoise
          NUS - Not usally suitable
          NA - Not applicable

140

Table 3.2

Q/ΔT W/K	COLD SIDE FLUID	PARAMETER	Low pressure Gas -1 bar	High pressure Gas -20 bar	Process Water	Low* viscosity organic fluid	High viscosity Liquid	Condensing Steam	Condensing Hydrocarbon	Condensing Hydrocarbon with Inert Gas
1 000	Low pressure gas (-1 bar)	U (W/m²K)	55	93	102	99	63	107	100	86
		C (£/(W/K))	4.36	3.76	3.43	3.54	4.29	3.27	3.50	3.14
	High pressure gas (-20 bar)	U (W/m²K)	93	300	429	375	120	530	388	240
		C (£/(W/K))	3.76	1.70	2.45	2.80	2.91	1.99	2.70	2.12
	Treated Cooling Water	U (W/m²K)	105	484	938	714	142	1607	764	345
		C (£/(W/K))	3.33	2.17	1.12	1.47	2.46	0.65	1.37	3.04
	Low viscosity organic liquid*	U (W/m²K)	99	375	600	500	130	818	524	286
		C (£/(W/K))	3.54	2.8	1.75	2.1	2.69	1.28	2.00	1.78
	High viscosity liquid**	U (W/m²K)	68	138	161	153	82	173	155	214
		C (£/(W/K))	3.97	2.54	3.16	3.33	3.29	2.94	3.28	2.82
	Boiling water	U (W/m²K)	105	467	875	677	140	1432	722	336
		C (£/(W/K))	3.33	2.25	1.20	1.55	2.50	0.73	1.45	3.13
	Boiling organic liquid***	U (W/m²K)	99	375	600	500	130	818	524	286
		C (£/(W/K))	3.53	2.8	1.75	2.1	2.69	1.28	2.00	1.78
5 000	Low pressure gas (-1 bar)	U (W/m²K)	55	93	102	99	63	107	100	86
		C (£/(W/K))	1.42	1.25	1.14	1.17	1.49	1.08	1.16	1.21
	High pressure gas (-20 bar)	U (W/m²K)	93	300	429	375	120	530	388	240
		C (£/(W/K))	1.25	0.8	0.63	0.72	1.12	0.66	0.70	0.77
	Treated Cooling Water	U (W/m²K)	105	484	938	720	142	1607	764	345
		C (£/(W/K))	1.10	0.72	0.54	0.71	0.94	0.32	0.67	0.78
	Low viscosity organic liquid*	U (W/m²K)	99	375	600	500	130	818	524	286
		C (£/(W/K))	1.17	0.72	0.58	0.70	1.03	0.62	0.67	0.84
	High viscosity liquid**	U (W/m²K)	68	138	161	153	82	173	155	124
		C (£/(W/K))	1.38	0.97	0.83	0.88	1.27	0.77	0.864	1.08
	Boiling water	U (W/m²K)	105	467	875	677	140	1432	722	336
		C (£/(W/K))	1.10	0.75	0.58	0.52	0.96	0.36	0.70	0.803
	Boiling organic liquid***	U (W/m²K)	99	375	600	500	130	818	524	286
		C (£/(W/K))	1.17	0.72	0.58	0.70	1.03	0.62	0.67	0.84
30 000	Low pressure gas (-1 bar)	U (W/m²K)	55	93	102	99	63	107	100	86
		C (£/(W/K))	0.82	0.54	0.52	0.53	0.73	0.49	0.52	0.58
	High pressure gas (-20 bar)	U (W/m²K)	93	300	500	375	120	530	388	240
		C (£/(W/K))	0.54	0.26	0.23	0.25	0.45	0.22	0.24	0.3
	Treated Cooling Water	U (W/m²K)	105	484	938	714	142	1607	764	345
		C (£/(W/K))	0.50	0.24	0.14	0.19	0.39	0.15	0.16	0.23
	Low viscosity organic liquid*	U (W/m²K)	99	375	600	500	130	818	524	286
		C (£/(W/K))	0.53	0.25	0.19	0.23	0.43	0.16	0.20	0.27
	High viscosity liquid**	U (W/m²K)	68	138	161	153	82	173	155	124
		C (£/(W/K))	0.69	0.41	0.40	0.34	0.59	0.37	0.36	0.44
	Boiling water	U (W/m²K)	105	467	875	677	140	1432	722	336
		C (£/(W/K))	0.50	0.22	0.15	0.17	0.4	0.13	0.19	0.23
	Boiling organic liquid***	U (W/m²K)	99	375	600	500	130	818	524	286
		C (£/(W/K))	0.53	0.25	0.19	0.21	0.43	0.16	0.20	0.27
100 000	Low pressure gas (-1 bar)	U (W/m²K)	55 #	93 #	102 #	99 #	63 #	107 #	100 #	86 #
		C (£/(W/K))	0.8	0.47	0.43	0.44	0.7	0.41	0.44	0.51
	High pressure gas (-20 bar)	U (W/m²K)	93 #	300	429	375	120 #	530	388	240
		C (£/(W/K))	0.47	0.17	0.13	0.14	0.37	0.11	0.14	0.20
	Treated Cooling Water	U (W/m²K)	105 #	484	938	714	142 #	1607	764	345
		C (£/(W/K))	0.42	0.12	0.083	0.095	0.31	0.065	0.094	0.015
	Low viscosity organic liquid*	U (W/m²K)	99 #	375	609	500	130 #	818	524	286
		C (£/(W/K))	0.44	0.14	0.11	0.11	0.348	0.088	0.11	0.17
	High viscosity liquid**	U (W/m²K)	68 #	138 #	161	153	82 #	173	155	124 #
		C (£/(W/K))	0.65	0.32	0.27	0.29	0.54	0.25	0.28	0.35
	Boiling water	U (W/m²K)	105 #	467	875	677	140 #	1432	722	336
		C (£/(W/K))	0.42	0.12	0.082	0.10	0.31	0.066	0.094	0.15
	Boiling organic liquid***	U (W/m²K)	99	375	600	500	130 #	818	524	286
		C (£/(W/K))	0.44	0.14	0.11	0.11	0.34	0.088	0.11	0.17
1000 000	Low pressure gas (-1 bar)	U (W/m²K)	55 #	93 #	102 #	99 #	63 #	107 #	100 #	86 #
		C (£/(W/K))	0.80	0.47	0.46	0.44	0.70	0.41	0.44	0.51
	High pressure gas (-20 bar)	U (W/m²K)	93 #	300 #	500 #	375 #	120 #	530 #	388 #	240 #
		C (£/(W/K))	0.47	0.15	0.088	0.12	0.37	0.0	0.11	0.18
	Treated Cooling Water	U (W/m²K)	105 #	484 #	938 #	714 #	142 #	1607	764 #	345 #
		C (£/(W/K))	0.42	0.091	0.047	0.062	0.31	0.027	0.058	0.13
	Low viscosity organic liquid*	U (W/m²K)	99 #	375 #	600 #	500 #	130 #	818 #	524 #	286 #
		C (£/(W/K))	0.44	0.12	0.073	0.088	0.34	0.054	0.084	0.15
	High viscosity liquid**	U (W/m²K)	68 #	138 #	161 #	153 #	82 #	173 #	155 #	124 #
		C (£/(W/K))	0.65	0.32	0.27	0.29	0.54	0.25	0.28	0.35
	Boiling water	U (W/m²K)	105 #	467 #	875 #	677 #	140 #	1432 #	722 #	336 #
		C (£/(W/K))	0.42	0.094	0.050	0.065	0.31	0.031	0.061	0.13
	Boiling organic liquid***	U (W/m²K)	99 #	375 #	600 #	500 #	130 #	818 #	524 #	286 #
		C (£/(W/K))	0.44	0.12	0.073	0.077	0.34	0.054	0.084	0.15

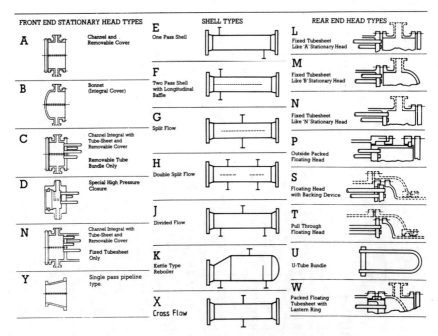

FRONT END STATIONARY HEAD TYPES	SHELL TYPES	REAR END HEAD TYPES

A — Channel and Removable Cover

B — Bonnet (Integral Cover)

C — Channel Integral with Tube-Sheet and Removable Cover / Removable Tube Bundle Only

D — Special High Pressure Closure

N — Channel Integral with Tube-Sheet and Removable Cover / Fixed Tubesheet Only

Y — Single pass pipeline type.

E — One Pass Shell

F — Two Pass Shell with Longitudinal Baffle

G — Split Flow

H — Double Split Flow

J — Divided Flow

K — Kettle Type Reboiler

X — Cross Flow

L — Fixed Tubesheet Like 'A' Stationary Head

M — Fixed Tubesheet Like 'B' Stationary Head

N — Fixed Tubesheet Like 'N' Stationary Head

P — Outside Packed Floating Head

S — Floating Head with Backing Device

T — Pull Through Floating Head

U — U-Tube Bundle

W — Packed Floating Tubesheet with Lantern Ring

Figure 3.4—Configuration nomenclature adopted by the Tubular Exchangers Manufacturers Association (TEMA)

The choice between various alternative TEMA types can be systematised in terms of a flow diagram as shown in Figure 3.5. Here, a series of questions is asked and the designer is directed towards specific designs depending on the answers to these questions. The questions are as follows.

(1) *Are there severe thermal expansion stresses?* Here the temperature differential between the shell and the tube metal temperature has to be calculated. The shell is generally at a temperature fairly close to the shell side fluid temperature and the tube metal is at a temperature intermediate between the tube side and shell side streams, its value depending on the heat transfer resistances in the two streams. The resulting differential thermal expansion is calculated and the strain induced in the shell-and-tubes estimated. The resultant stresses are evaluated against allowable stresses for the particular materials. Care has to be exercised to ensure that not only steady-state conditions are evaluated but also those which are likely to occur in startup and shutdown or in any other plant transient.

(2) *Are bellows allowed?* Sometimes the use of bellows is ruled out because of standard practices within the particular company using the heat exchanger, or due to pressure or temperature considerations. Bellows are certainly a source of failure in heat exchangers and their utilisation has to be assessed in the context of the particular plant. In the majority of cases, bellows are permissible.

(3) *Is there a high shell side fouling factor?* If the fouling factors are greater than,

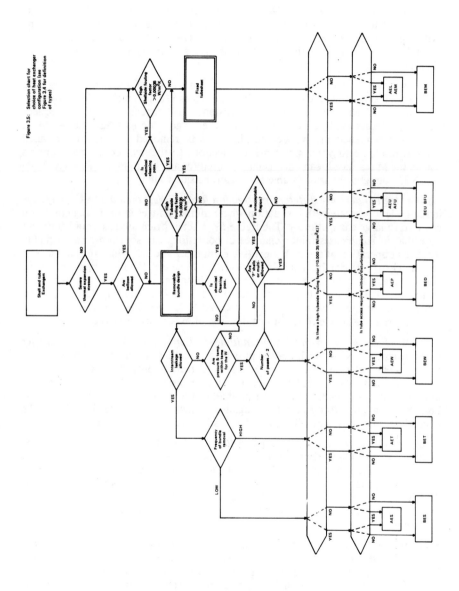

Figure 3.5—Selection chart for choice of heat exchanger configuration (see Figure 3.4 for definition of types)

143

say, 0.00035 W/m²K, shell side cleaning is very likely to be necessary on a periodic basis.

(4) *Is chemical cleaning possible?* Chemical cleaning is necessary for fixed tube sheet design for removal of fouling layers on the shell side. It is feasible with some hydrocarbons and for some water-formed scale. The question of chemical cleaning feasibility is relevant to both tube side and shell side cleaning as is seen from the diagram.

(5) *Is there a high tube side fouling factor?* This is a similar question to that arising for the shell side, though cleaning of the tube side is usually much easier. However, the choice of fixed tube sheet design, for instance, will depend on whether frequent cleaning is necessary. If it is necessary, it is better to go to designs AEL and AEM (in the TEMA nomenclature) where end plates can be removed and the tubes cleaned without disconnecting the pipe work connections to the heat exchanger. For this reason the M rear heads can only usually be used for an even number of passes. NEN and NEM designs are also suitable for high tube side fouling duties.

(6) *Is the temperature difference correction factor F_T less than 0.8?* Usually, it is inadvisable to operate with a situation where minor changes in operating condition can lead to large changes in F_T. The figure $F_T = 0.8$ is often used as a guideline, but note that in some cases minor changes in conditions can sometimes cause highly inefficient operation of the exchanger (see Figures 3.2 and 3.3) which may in fact fail completely to perform the required duty even for $F_T > 0.8$. U-tube designs are inevitably of at least two passes (by definition) and this limits their use to cases where F_T is greater than 0.8. It should be pointed out that the figure of 0.8 is somewhat subjective, some people suggest a somewhat lower figure, 0.75, say. However, a figure of 0.8 seems to be a reasonable consensus of the view of experienced designers.

(7) *Are F-shells (see Figure 3.4) or multiple shells in series allowed?* If the answer to this question is yes, then it may be possible to secure a sufficient increase in the F_T factor to allow the use of U-tube shells. If not, then the only alternative is to go to single pass units and since the choice has already been made to go to a removable bundle shell this implies the use of floating head units as is indicated in the chart. Sometimes, a very strong preference against F-shells and multiple shells has to be overridden since it is impossible to find a floating head unit of suitable design.

(8) *Will interstream leakage be hazardous?* All shell-and-tube heat exchangers are designed to separate the two process streams but interstage leakage may possibly occur due to breakdown of seals *etc.* If such interstage leakage would be a considerable hazard, the design configuration chosen has to be such as to minimise the chance of such leakage. For floating head designs, this implies the use of rear end head types P and W (Figure 3.4) where any leakage through the gasketed joint would be to atmosphere.

(9) *If rear end types P or W is to be used, are the pressure and temperature limitations within the limits specified by TEMA?* These limitations arise because of the limits on the integrity of the packed gland on the shell side of the P-type and on both sides of the W-type. The normal limits are around 20 atmosphere pressure and 200°C temperature. Note that for conditions exceeding these limits it may be necessary to reconsider some of the decisions made earlier in the selection process.

(10) *Is the number of passes required greater than two?*   If the tube side velocity considerations dictate the use of more than two passes, then this limits the choice for a floating head design to the P-type rear end head design.

(11) *Is the frequency of bundle removal low or high?*   This question relates to the most common form of floating head unit where a minor amount of interstream leakage is not too hazardous. If the frequency of bundle removal is large (say more than twice a year) then it is advisable to use the T-type (pull through floating head) rather than the S-type with a backing device. All floating head devices imply a significant bypass area between the bundle and the shell. This is particularly true in the case of the T-type design and it is usually necessary in this and in other floating head designs to insert sealing strips to minimise the effect of bypass on the heat transfer.

(12) *Is there a high tube side fouling factor?*   This is a similar question to that asked in the context of the fixed tube sheet (question 5 above).

(13) *Is tube end access required?*   If there is a high tube side fouling factor and *chemical cleaning* is not possible, then end access will be required which implies the use of a removable end plate, namely an A-type front end head design. Such a design may also be required if access to the tube ends was needed for other reasons (*e.g.* maintenance and inspection).

It will be noted that the selection procedure shown in Figure 3.5 leads only to a limited number of potential choices when compared with the alternatives shown in Figure 3.4. Specifically, front end head types A and B, shell type E and all rear end types except N are mentioned in the chart. The following comments can be made on the remaining alternatives.

(1) *Front end types*—Type C is not often used other than for high pressures (typically in excess of 100 bar). For very high pressures (typically in excess of 150 bar) it may be necessary to go to the special high pressure closure D. Front end type N can sometimes be more economical than type A, though there are sometimes difficulties in manufacturing and maintenance. Type Y is used when the exchanger is to be inserted in a pipeline, and minimises piping costs; it is sometimes known as a "cone-type" head. It is limited to a single pass on the tube side, or, with suitable partitioning, any odd number of tube side passes.

(2) *Shell types*—Types G, H, J and X are often used in condensers and where there is a requirement to minimise pressure drop. Type K is for kettle reboilers.

(3) *The remaining rear-end head type (N)* is only used when the front end is of that type. It may sometimes be economical, though it can provide difficulties in access and maintenance.

A design which has become popular in recent years is the so-called *rod-baffle* type in which the tubes are supported by arrays of parallel rods. These rod arrays are placed at intervals along the tube bundle, each alternate array being rotated by 90°. This design has the primary advantage of being extremely stable against tube vibration but, over limited ranges, it also gives a higher heat transfer coefficient for a given pressure drop. The disadvantages are that, for a given tube bundle, there is a limit on the *maximum achievable heat transfer coefficient* and there is also a problem in ensuring good flow distribution in the axial-type flow occurring, particularly near the inlet and outlet.

Table 3.3—Film coefficient used for shell-and-tube heat exchangers

	Cold side			Hot side		
	Film coefficient W/m²K (clean)	Fouling factor Km²/W	Overall film coefficient W/m²K	Film coefficient W/m²K (clean)	Fouling factor Km²/W	Overall film coefficient W/m²K
Low pressure gas ~ 1 bar	112	0.0002	110	112	0.0002	110
High pressure gas ~ 20 bar	682	0.0002	600	682	0.0002	600
Process water	—	—	—	6000	0.0005	1500
Treated cooling water	5000	0.0002	2500	—	—	—
Low viscosity organic liquid	1667	0.0004	1000	1667	0.0004	1000
High viscosity liquid	210	0.0008	180	170	0.0008	150
Condensing steam	—	—	—	8182	0.0001	4500
Condensing hydrocarbon	—	—	—	1410	0.0002	1100
Condensing hydrocarbon/1 bar	—	—	—	435	0.0002	400
Boiling treated water	5676	0.0003	2100	—	—	—
Boiling organic liquid	1667	0.0004	1000	—	—	—

In order to provide a basis for quick estimation of size and comparative cost, both coefficient and cost data are required. For shell-and-tube head exchangers, it was decided to estimate, on the basis of experience, the types of coefficient which would occur for a variety of fluids as shown in Table 3.3. For a given combination, the reciprocal of the overall coefficient U was calculated as the sum of the reciprocals of the hot and cold side overall film coefficients (including fouling). The combined U values are given for the various fluid pairs in Table 3.2.

The costs of exchangers vary from manufacturer to manufacturer and the specification of cost data is therefore difficult. However, to substantiate the present exercise, a very large number of cost estimates were made covering a whole range of systems and designs and the resultant data is presented in Figure 3.6. Figure 3.6a shows the outside surface area as a function of shell diameter and length. The full line shows the area used in preparation of Table 3.2. Figure 3.6 shows the variation of price per square meter as a function of shell diameter and length. The data shown in Figure 3.6b are for the "standard" case of a shell-and-tube heat exchanger operating at less than 10 bar pressure and constructed from carbon steel with a fixed tube plate construction (type BEM). Figure 3.6c and 3.6d show correction factors for the more common alternative construction types. Figures 3.6e and 3.6f show correction factors for higher pressures and Figures 3.6g and 3.6h show correction factors to be used if stainless steel is employed as the construction material. The base price should be multiplied by the three correction factors in a cumulative manner.

Using the same methods as those which have been used to generate the curves shown in Figure 3.6b (BEM, carbon steel), C values were estimated and are included for all the fluid pairs and for three values of $Q/\Delta T$ in Table 3.2.

3.4.2. Double-pipe heat exchangers
Double-pipe heat exchangers may be of various forms as illustrated in the insets in Tables 3.4 and 3.5. They can be in the form of a single tube within an outer tube or as a group of tubes as illustrated. Double-pipe heat exchangers have the advantage that they offer almost perfect countercurrent flow. For shell sizes of 150mm and above,

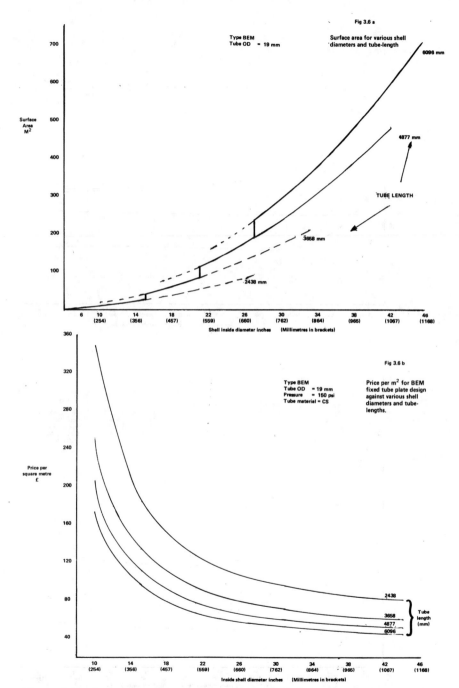

Figure 3.6—Cost curves for shell-and-tube heat exchangers (*courtesy of Johnson Hunt Ltd*)

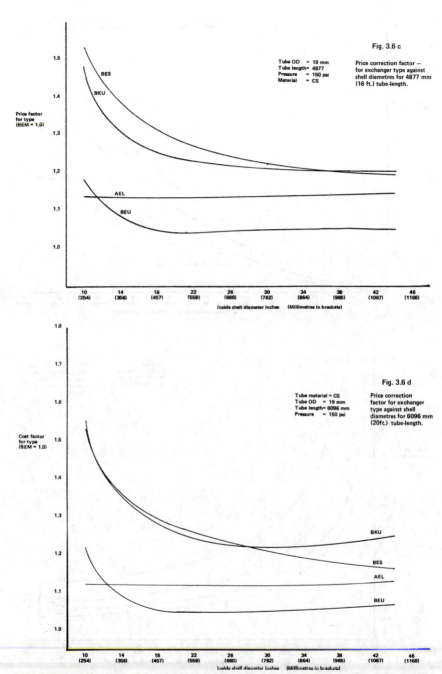

Figure 3.6—Cost curves for shell-and-tube heat exchangers (*courtesy of Johnson Hunt Ltd*)

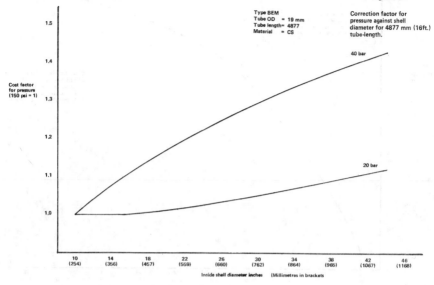

Fig. 3.6 e

Type BEM
Tube OD   = 19 mm
Tube length= 4877
Material   = CS

Correction factor for
pressure against shell
diameter for 4877 mm (16ft.)
tube-length.

40 bar

20 bar

Cost factor
for pressure
(150 psi = 1)

1.5

1.4

1.3

1.2

1.1

1.0

10
(254)
14
(356)
18
(457)
22
(559)
26
(660)
30
(762)
34
(864)
38
(965)
42
(1067)
46
(1168)

Inside shell diameter inches   (Millimetres in brackets)

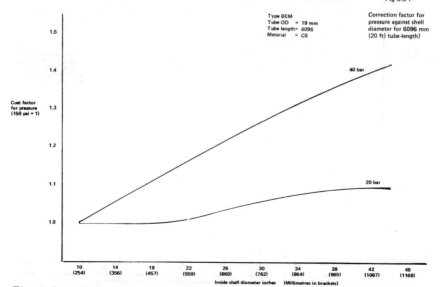

Fig 3.6 f

Type BEM
Tube OD   = 19 mm
Tube length= 6096
Material   = CS

Correction factor for
pressure against shell
diameter for 6096 mm
(20 ft) tube-length)

40 bar

20 bar

Cost factor
for pressure
(150 psi = 1)

1.5

1.4

1.3

1.2

1.1

1.0

10
(254)
14
(356)
18
(457)
22
(559)
26
(660)
30
(762)
34
(864)
38
(965)
42
(1067)
46
(1168)

Inside shell diameter inches   (Millimetres in brackets)

Figure 3.6—Cost curves for shell-and-tube heat exchangers (*courtesy of Johnson Hunt Ltd*)

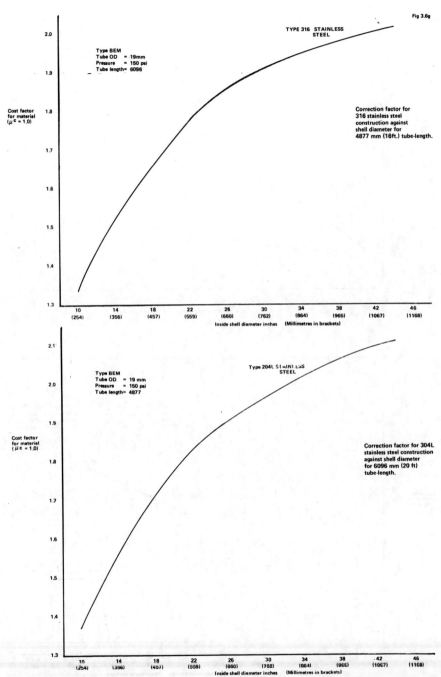

Figure 3.6—Cost curves for shell-and-tube heat exchangers (*courtesy of Johnson Hunt Ltd*)

segmental baffles are commonly employed and these units become rather similar to a single pass counter flow shell-and-tube heat exchanger. They are sometimes described as jacketed U-tube exchangers. For shells with sizes less than 150mm, there are no baffles, but the tubes are supported using special spider supports. In finned double-pipe exchangers, baffles are not necessary since the tubes are inherently self-supporting within the unit. Intermeshing of the fins is prevented by wrapping strips of stainless steel or other suitable material around the outside of the fins. A further stainless steel band is wound round the outside of the bundle to hold the bundle together to allow for ease of construction. With both the plain tube and fin tube multitubular designs, peripheral baffles may be necessary to block the leakage area between the bundle and the shell. As will be seen, if axial flow is maintained in the units, then the problems of mixing and leakage associated with baffled units can be avoided. This makes the axial flow units (without baffles) particularly attractive for heat transfer at low Reynolds numbers.

The choice between plain and finned double-pipe heat exchangers is usually made ultimately on the basis of cost for a particular duty. In general, it is found, that the finned tube type is more economic where the heat transfer coefficients are relatively high on the tube side and where there is a need to enhance the surface on the shell side coefficient. Roughly speaking, the rule is that such units are economic when the outside coefficient is less than half the inside coefficient. U and C values for double-pipe heat exchangers are given in Tables 3.4 and 3.5. The U value for the finned tube case is based on the *total* extended surface area on the outside of the tube. For the larger duties, some of the cost figures are for multiple units. In double-pipe heat exchangers, the penalty for using multiple units is much less than for other forms of heat exchanger, due to their modular construction. They are specifically designed to bolt together to form whichever configuration is most appropriate for the particular duty. This facility is illustrated in Figure 3.7.

In contrast to shell-and-tube exchangers where the TEMA nomenclature has been adapted as the standard, there is no internationally accepted standard nomenclature for double-pipe units. However, the nomenclature adapted by Brown Fintube has achieved some international currency and this nomenclature is defined in Table 3.6.

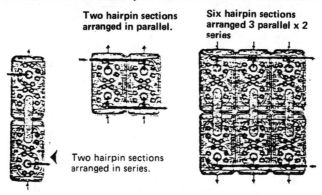

Figure 3.7—Arrangement of sections in a double-pipe heat exchanger (*courtesy of Filtration and Transfer Ltd*)

Table 3.4 Courtesy of Filtration and Transfer Ltd.

Type: Double Pipe — Bare Tube

Diagrams and Comments

Single Tube (0.3 to 10 m²)

Movable brackets

Multitubular (4", 6" 8" N/B Shells — 3.0 to 45 m²)

Movable brackets

Multitubular (10", 12" & 16" N/B Shells — 30 m² +)

Movable brackets

Notes:
*   Viscosity range 1 – 5 centipoise
**   Viscosity range > 100 centipoise
*** Viscosity typically < 1 centipoise
NUS - Not usually suitable - try fintube double pipe.
NA - Not applicable

Q/ΔT w/k	COLD SIDE FLUID	PARAMETER	Low Pressure Gas ~1 Bar	High Pressure Gas ~20 Bar	Process Water	Low Viscosity Organic Fluid	High Viscosity Fluid	Cohdensing Steam	Condensing Hydrocarbon	Condensing Hydrocarbon with Inert Gas
1 000	Low pressure gas (~1 bar)	U (W/m^2 K)	55	93	102	99	63	107	100	85
		C (£/(W/K))	3.182	1.882	1.716	1.768	2.778	1.636	1.750	2.059
	High pressure gas (~20 bar)	U (W/m^2 K)	93	300	429	375	120	530	388	240
		C (£/(W/K))	1.882	0.583	1.515	1.733	1.458	1.226	1.675	0.729
	Treated Cooling Water	U (W/m^2 K)	105	484	938	714	142	1607	764	345
		C (£/(W/K))	1.667	1.343	0.693	0.910	1.232	0.404	0.851	1.884
	Low viscosity organic liquid*	U (W/m^2 K)	99	375	600	500	130	818	524	286
		C (£/(W/K))	1.768	1.733	1.083	1.300	1.346	0.795	1.240	0.612
	High viscosity liquid**	U (W/m^2 K)	68	138	161	153	82	173	155	124
		C (£/(W/K))	2.574	1.268	1.087	1.144	2.134	1.012	1.129	1.411
	Boiling water	U (W/m^{2}K)	105	467	875	677	140	1432	722	336
		C (£/(W/K))	1.667	1.392	0.743	0.960	1.250	0.454	0.900	1.935
	Boiling organic liquid***	U (W/m^{2}K)	99	375	600	500	130	818	524	286
		C (£/(W/K))	1.768	1.733	1.083	1.300	1.346	0.795	1.240	0.612
5 000	Low pressure gas (~1 bar)	U (W/m^2 K)	55	93	102	99	63	107	100	85
		C (£/(W/K))	2.364	1.398	1.275	1.313	2.063	1.215	1.300	1.529
	High pressure gas (~20 bar)	U (W/m^2 K)	93	300	429	375	120	530	388	240
		C (£/(W/K))	1.398	0.583	0.408	0.467	1.083	0.330	0.451	0.729
	Water	U (W/m^2 K)	105	484	938	714	142	1607	764	345
		C (£/(W/K))	1.238	0.362	0.187	0.245	1.232	0.109	0.229	0.507
	Low viscosity organic liquid*	U (W/m^2 K)	99	375	600	500	130	818	524	286
		C (£/(W/K))	1.313	0.467	0.292	0.350	1.346	0.214	0.334	0.612
	High viscosity liquid**	U (W/m^2 K)	68	138	161	153	82	173	155	124
		C (£/(W/K))	1.912	1.268	1.087	1.144	1.585	0.994	1.129	1.411
	Boiling water	U (W/m^2 K)	105	467	875	677	140	1432	722	336
		C (£/(W/K))	1.238	0.374	0.200	0.258	1.250	0.122	0.242	0.521
	Boiling organic liquid	U (W/m^2 K)	99	375	600	500	130	818	524	286
		C (£/(W/K))	1.313	0.467	0.292	0.350	1.346	0.214	0.334	0.612
30 000	Low pressure gas (~1 bar)	U (W/m^2 K)	55	93	102	99	63	107	100	85
		C (£/(W/K))	2.364	1.398	1.275	1.313	2.063	1.215	1.300	1.529
	High pressure gas (~20 bar)	U (W/m^2 K)	93	300	429	375	120	530	388	240
		C (£/(W/K))	1.398	0.433	0.303	0.347	1.083	0.245	0.335	0.542
	Treated Cooling Water	U (W/m^2 K)	105	484	938	714	142	1607	764	345
		C (£/(W/K))	1.238	0.269	0.187	0.182	0.915	0.109	0.229	0.377
	Low viscosity organic liquid*	U (W/m^2 K)	99	375	600	500	130	818	524	286
		C (£/(W/K))	1.313	0.347	0.217	0.260	1.00	0.214	0.248	0.455
	High viscosity liquid**	U (W/m^2 K)	68	138	161	153	82	173	155	124
		C (£/(W/K))	1.912	0.942	0.807	0.850	1.585	0.751	0.839	1.048
	Boiling water	U (W/m^2 K)	105	467	875	677	140	1432	722	336
		C (£/(W/K))	1.238	0.278	0.200	0.192	0.929	0.122	0.180	0.387
	Boiling organic liquid***	U (W/m^2 K)	99	375	600	500	130	818	524	286
		C (£/(W/K))	1.313	0.347	0.217	0.260	1.0	0.214	0.248	0.455
100 000	Low pressure gas (~1 bar)	U (W/m^2 K)	55	93	102	99	63	107	100	85
		C (£/(W/K))	2.364	1.398	1.275	1.313	2.063	1.215	1.300	1.529
	High pressure gas (~20 bar)	U (W/m^2 K)	93	300	429	375	120	530	388	240
		C (£/(W/K))	1.398	0.433	0.303	0.347	1.083	0.245	0.335	0.542
	Treated Cooling Water	U (W/m^{2}K)	105	484	938	714	142	1607	764	345
		C (£/(W/K))	1.238	0.269	0.139	0.182	0.915	0.081	0.170	0.377
	Low viscosity organic liquid*	U (W/m^{2}K)	99	375	600	500	130	818	524	286
		C (£/(W/K))	1.313	0.347	0.217	0.260	1.00	0.159	0.248	0.455
	High viscosity liquid**	U (W/m^{2}K)	68	138	161	153	82	173	155	124
		C (£/(W/K))	1.912	0.942	0.807	0.850	1.585	0.751	0.839	1.048
	Boiling water	U (W/m^2 K)	105	467	875	677	140	1432	722	336
		C (£/(W/K))	1.238	0.278	0.149	0.192	0.929	0.091	0.180	0.387
	Boiling organic liquid	U (W/m^2 K)	99	375	600	500	130	818	524	286
		C (£/(W/K))	1.313	0.347	0.217	0.260	1.00	0.159	0.248	0.455
1 000 000	Low pressure gas (~1 bar)	U (W/m^{2}K)	55	93	102	99	63	107	100	85
		C (£/(W/K))	2.364	1.398	1.275	1.313	2.063	1.215	1.300	1.529
	High pressure gas (~20 bar)	U (W/m^{2}K)	93	300	429	375	120	530	388	240
		C (£/(W/K))	1.398	0.433	0.303	0.347	1.083	0.245	0.335	0.542
	Treated Cooling Water	U (W/m^{2}K)	105	484	938	714	142	1607	764	345
		C (£/(W/K))	1.238	0.269	0.139	0.182	0.915	0.081	0.170	0.377
	Low viscosity organic liquid*	U (W/m^{2}K)	99	375	600	500	130	818	524	286
		C (£/(W/K))	1.313	0.347	0.217	0.260	1.00	0.159	0.248	0.455
	High viscosity liquid**	U (W/m^{2}K)	68	138	161	153	82	173	155	124
		C (£/(W/K))	1.912	0.942	0.807	0.850	1.585	0.751	0.839	1.048
	Boiling water	U (W/m^{2}K)	105	467	875	677	140	1432	722	336
		C (£/(W/K))	1.238	0.278	0.149	0.192	0.929	0.091	0.180	0.387
	Boiling organic liquid	U (W/m^{2}K)	99	375	600	500	130	818	524	286
		C (£/(W/K))	1.313	0.347	0.217	0.260	1.0	0.159	0.248	0.455

Table 3.5 Courtesy of Filtration and Transfer Ltd.

Type: Double Pipe — with longitudinal fins

Diagrams and Comments

Single Fintube: — Subscript 'a'

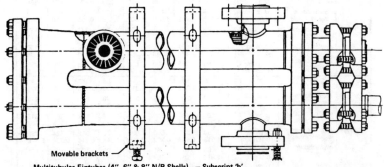

Movable brackets

Multitubular Fintubes (4″, 6″ & 8″ N/B Shells) — Subscript 'b'

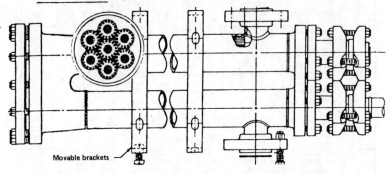

Movable brackets

Multitubular Fintubes (10″ & 12″ N/B Shells) — Subscript 'c'

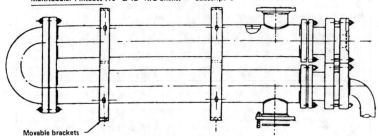

Movable brackets

Notes: *   Viscosity range 1 – 5 centipoise
           **   Viscosity range > 100 centipoise
           *** Viscosity typically < 1 centipoise

Lower film coefficient fluid shellside.

HOT SIDE FLUID

Q/ΔT	COLD SIDE FLUID	PARAMETER	Low pressure Gas ~1 Bar	High Pressure Gas ~20 Bar	Process Water	Low viscosity Organic Fluid	High viscosity Liquid	Condensing Steam	Condensing Hydro-carbon	Condensing Hydro-Carbon with Inert Gas
W/K	Low pressure gas (~1 bar)	U (W/m² K)	21.0	56.0	77.0	68.0	23.0	70.0	69.0	42.0
		C (£/(W/K))	2.4c	1.4b	1.0b	1.2b	2.2c	1.1b	1.2b	1.9b
	High pressure gas (~20 bar)	U (W/m² K)	56.0	96.5	181.0	136.0	70.0	182.0	153.0	91.0
		C (£/(W/K))	1.4b	0.83b	0.44b	0.59b	1.1b	0.66a	0.52b	0.88b
	Treated Cooling Water	U (W/m² K)	87.0	148.0	182.0	176.0	108.0	261.0	176.0	128.0
		C (£/(W/K))	0.92b	0.81a	0.66a	0.68a	0.74b	0.46a	0.68a	0.94a
1 000	Low viscosity organic liquid*	U (W/m² K)	68.0	136.0	199.0	153.0	80.5	221.0	168.0	125.0
		C (£/(W/K))	1.2b	0.59b	0.40b	0.52b	0.99b	0.54a	0.48b	0.64b
	High viscosity liquid**	U (W/m² K)	24.5	68.0	105.0	91.0	25.0	139.0	94.0	50.5
		C (£/(W/K))	2.0c	1.2b	0.76b	0.88b	2.0c	0.58b	0.85b	1.6b
	Boiling water	U (W/m² K)	84.0	133.0	159.0	170.0	140.0	260.0	153.0	187.0
		C (£/(W/K))	0.95b	0.90a	0.75a	0.47b	0.57b	0.46a	0.78a	0.43b
	Boiling organic liquid***	U (W/m² K)	68.0	136.0	199.0	153.0	80.5	221.0	168.0	125.0
		C (£/(W/K))	1.2b	0.59b	0.40b	0.52b	0.99b	0.54a	0.48b	0.64b
	Low pressure gas (~1 bar)	U (W/m² K)	21.0	56.0	77.0	68.0	23.0	70.0	69.0	42.0
		C (£/(W/K))	2.38c	0.89c	0.65c	0.74c	2.17c	0.71c	0.72c	1.19c
	High pressure gas (~20 bar)	U (W/m² K)	56.0	96.5	181.0	136.0	70.0	301.0	153.0	91.0
		C (£/(W/K))	0.89c	0.52c	0.44b	0.37c	0.71c	0.27b	0.33c	0.55c
	Water	U (W/m² K)	87.0	238.5	323.5	295.0	108.0	534.0	307.0	204.5
		C (£/(W/K))	0.57c	0.34b	0.25b	0.27b	0.46c	0.15b	0.26b	0.39b
5 000	Low viscosity organic liquid*	U (W/m² K)	68.0	136.0	199.0	153.0	80.5	375.0	168.0	125.0
		C (£/(W/K))	0.74c	0.37c	0.40b	0.33c	0.62c	0.21b	0.48b	0.40c
	High viscosity liquid**	U (W/m² K)	24.5	68.0	105.0	91.0	25.0	139.0	94.0	50.5
		C (£/(W/K))	2.05c	0.74c	0.48c	0.55c	2.0c	0.36c	0.53c	0.99c
	Boiling water	U (W/m² K)	84.0	216.0	284.0	170.0	140.0	511.0	267.0	187.0
		C (£/(W/K))	0.60c	0.37b	0.28b	0.47b	0.36c	0.16b	0.30b	0.43b
	Boiling organic liquid***	U (W/m² K)	68.0	136.0	199.0	153.0	80.5	375.0	168.0	125.0
		C (£/(W/K))	0.74c	0.37c	0.40b	0.33c	0.62c	0.21b	0.48b	0.40c
	Low pressure gas (~1 bar)	U (W/m² K)	21.0	56.0	77.0	68.0	23.0	70.0	69.0	42.0
		C (£/(W/K))	1.9c	0.80c	0.58c	0.66c	1.7c	0.64c	0.65c	1.1c
	High pressure gas (~20 bar)	U (W/m² K)	56.0	96.5	181.0	136.0	70.0	301.0	153.0	91.0
		C (£/(W/K))	0.80c	0.47c	0.28c	0.37c	0.64c	0.17c	0.33c	0.49c
	Treated Cooling Water	U (W/m² K)	87.0	238.5	323.5	295.0	108.0	534.0	307.0	204.5
		C (£/(W/K))	0.52c	0.21c	0.15c	0.17c	0.42c	0.094c	0.16c	0.24c
30 000	Low viscosity organic liquid*	U (W/m² K)	68.0	136.0	199.0	153.0	80.5	375.0	168.0	125.0
		C (£/(W/K))	0.66c	0.37c	0.25c	0.33c	0.56c	0.13c	0.30c	0.40c
	High viscosity liquid**	U (W/m² K)	24.5	68.0	105.0	91.0	25.0	139.0	94.0	50.5
		C (£/(W/K))	1.6c	0.66c	0.43c	0.49c	1.6c	0.36c	0.48c	0.89c
	Boiling water	U (W/m² K)	84.0	216.0	284.0	170.0	140.0	511.0	267.0	187.0
		C (£/(W/K))	0.54c	0.23c	0.18c	0.29c	0.36c	0.098c	0.19c	0.27c
	Boiling organic liquid***	U (W/m² K)	68.0	136.0	199.0	153.0	80.5	375.0	168.0	125.0
		C (£/(W/K))	0.66c	0.37c	0.25c	0.33c	0.56c	0.13c	0.30c	0.4c
	Low pressure gas (~1 bar)	U (W/m² K)	21.0	56.0	77.0	68.0	23.0	90.0	69.0	42.0
		C (£/(W/K))	1.9c	0.71c	0.52c	0.59c	1.7c	0.44c	0.58c	0.95c
	High pressure gas (~20 bar)	U (W/m² K)	56.0	96.5	181.0	136.0	70.0	301.0	153.0	91.0
		C (£/(W/K))	0.71c	0.41c	0.25c	0.33c	0.57c	0.15c	0.29c	0.44c
	Water	U (W/m² K)	87.0	238.5	325.5	295.0	108.0	534.0	307.0	204.5
		C (£/(W/K))	0.46c	0.19c	0.14c	0.15c	0.42c	0.094c	0.15c	0.22c
100 000	Low viscosity organic liquid*	U (W/m² K)	68.0	136.0	199.0	153.0	80.5	375.0	168.0	125.0
		C (£/(W/K))	0.59c	0.33c	0.23c	0.29c	0.50c	0.12c	0.27c	0.36c
	High viscosity liquid**	U (W/m² K)	24.5	68.0	105.0	91.0	25.0	134.0	94.0	50.5
		C (£/(W/K))	1.6c	0.59c	0.43c	0.44c	1.6c	0.34c	0.43c	0.79c
	Boiling water	U (W/m² K)	84.0	216.0	284.0	170.0	19.0	511.0	267.0	187.0
		C (£/(W/K))	0.48c	0.21c	0.16c	0.26c	2.1c	0.098c	0.17c	0.24c
	Boiling organic liquid***	U (W/m² K)	68.0	136.0	199.0	153.0	80.5	375.0	168.0	125.0
		C (£/(W/K))	0.59c	0.33c	0.23c	0.29c	0.50c	0.12c	0.27c	0.36c
	Low pressure gas (~1 bar)	U (W/m² K)	21.0	56.0	77.0	68.0	23.0	90.0	69.0	42.0
		C (£/(W/K))	1.9c	0.71c	0.52c	0.59c	1.7c	0.44c	0.58c	0.95c
	High pressure gas (~20 bar)	U (W/m² K)	56.0	96.5	181.0	136.0	70.0	301.0	153.0	91.0
		C (£/(W/K))	0.71c	0.41c	0.22c	0.29c	0.57c	0.13c	0.26c	0.44c
	Treated Cooling Water	U (W/m² K)	87.0	238.5	325.5	295.0	108.0	534.0	307.0	204.5
		C (£/(W/K))	0.46c	0.17c	0.12c	0.14c	0.37c	0.075c	0.13c	0.20c
1 000 000	Low viscosity organic liquid*	U (W/m² K)	68.0	136.0	199.0	153.0	80.5	375.0	168.0	125.0
		C (£/(W/K))	0.59c	0.29c	0.20c	0.26c	0.50c	0.11c	0.24c	0.32c
	High viscosity liquid**	U (W/m² K)	24.5	68.0	105.0	91.0	25.0	134.0	94.0	50.5
		C (£/(W/K))	1.6c	0.59c	0.38c	0.44c	1.6c	0.30c	0.43c	0.79c
	Boiling water	U (W/m² K)	84.0	216.0	284.0	170.0	19.0	511.0	267.0	187.0
		C (£/(W/K))	0.48c	0.19c	0.14c	0.24c	2.1c	0.078c	0.15c	0.21c
	Boiling organic liquid	U (W/m² K)	68.0	136.0	199.0	153.0	80.5	375.0	168.0	125.0
		C (£/(W/K))	0.59c	0.29c	0.20c	0.26c	0.50c	0.11c	0.24c	0.32c

155

Table 3.6—Standard Hairpin Exchanger Nomenclature
(courtesy of Filtration and Transfer Limited, Brown Fintube licensee in the UK)

Section type	No. tubes	Shell O.D.	Tube O.D.	Shell nozzle (2) Ansi. WN.R.F.	Tube nozzle (1) (2)	Height Low Press	Height High Press	Bracket Width Low Press	Bracket Width High Press	Overall length add to nominal fin length	Type of baffle or support
51	1	3.5"	1.9"	2"	1½" B.W.	13½"	14½"	8"	9"	1'6"	—
53	1	4.5"	1.9"	3"	1½" B.W.	16"	18"	10"	10"	1'6"	—
54	1	4.5"	2.875"	3"	2½" B.W.	16"	18"	10"	10"	1'6"	—
55	7	4.5"	0.75"	3"	2½" S.W.	16" (4)	16" (4)	10"	10"	1'8"	Peripheral
56	7	4.5"	0.875"	3"	2½" S.W.	16"	—	10"	10"	1'8"	—
57	7(3)	4.5"	0.875"	3"	2½" S.W.	16" (4)	16" (4)	10"	10"	1'8"	Spider
58	7(3)	4.5"	1.0"	3"	2½" S.W.	16"	—	10"	—	1'8"	Spider
61	19	8.625"0.75"		6"	4" S.W.	30"	(5)	14"	(5)	2'4"	Peripheral
62	19	8.625"0.875"		6"	4" S.W.	30"	(5)	14"	(5)	2'4"	Peripheral
64	7	6.625"1.0"		4"	3" S.W.	24½"	(5)	14"	(5)	1'11"	—
65	14	6.625"0.75"		4"	3" S.W.	24½"	(5)	14"	(5)	1'11"	Peripheral
66	22(3)	6.625"0.75"		4"	3" S.W.	24½"	(5)	14"	(5)	1'11"	Segmental
67	22(3)	6.625"0.75"		4"	3" S.W.	24½"	(5)	14"	(5)	1'11"	Support
70	55(3)	8.625"0.625"		6"	4" S.W.	30"	(5)	14"	(5)	2'4"	Segmental
71	37(3)	8.625"0.75"		6"	4" S.W.	30"	(5)	14"	(5)	2'4"	Segmental
72	31(3)	8.625"0.875"		6"	4" S.W.	30"	(5)	14"	(5)	2'4"	Segmental
73	55(3)	8.625"0.625"		6"	4" S.W.	30"	(5)	14"	(5)	2'4"	Support
74	37(3)	8.625"0.75"		6"	4" S.W.	30"	(5)	14"	(5)	2'4"	Support
75	31(3)	8.625"0.875"		6"	4" S.W.	30"	(5)	14"	(5)	2'4"	Support
80	1	2.375"1.0"		1½"	¾" B.W.	12"	(5)	7"	(5)	1'6"	—
81	1	3.5" 1.0"		2"	¾" B.W.	12"	(5)	7"	(5)	1'6"	—

B.W. = Beveled for welding   S.W. = Socket weld
(1) Flanged spool pieces can be supplied on request.
(2) Flange rating to suit design conditions.

(3) Available with bare tubes only.
(4) When tubeside design pressure over 1800 psi add 2½".
(5) Available on application.

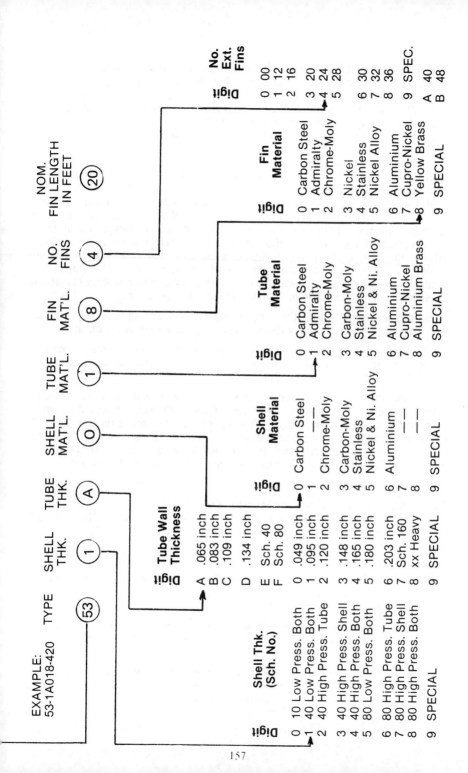

EXAMPLE:
53-1A018-420

TYPE	SHELL THK.	TUBE THK.	SHELL MAT'L.	TUBE MAT'L.	FIN MAT'L.	NO. FINS	NOM. FIN LENGTH IN FEET
(53)	(1)	(A)	(0)	(1)	(8)	(4)	(20)

Shell Thk. (Sch. No.)

Digit	
0	10 Low Press. Both
1	40 Low Press. Both
2	40 High Press. Tube
3	40 High Press. Shell
4	40 High Press. Both
5	80 Low Press. Both
6	80 High Press. Tube
7	80 High Press. Shell
8	80 High Press. Both
9	SPECIAL

Tube Wall Thickness

Digit	
A	.065 inch
B	.083 inch
C	.109 inch
D	.134 inch
E	Sch. 40
F	Sch. 80
0	.049 inch
1	.095 inch
2	.120 inch
3	.148 inch
4	.165 inch
5	.180 inch
6	.203 inch
7	Sch. 160
8	xx Heavy
9	SPECIAL

Shell Material

Digit	
0	Carbon Steel
1	—
2	Chrome-Moly
3	Carbon-Moly
4	Stainless
5	Nickel & Ni. Alloy
6	Aluminium
7	—
8	—
9	SPECIAL

Tube Material

Digit	
0	Carbon Steel
1	Admiralty
2	Chrome-Moly
3	Carbon-Moly
4	Stainless
5	Nickel & Ni. Alloy
6	Aluminium
7	Cupro-Nickel
8	Aluminium Brass
9	SPECIAL

Fin Material

Digit	
0	Carbon Steel
1	Admiralty
2	Chrome-Moly
3	Nickel
4	Stainless
5	Nickel Alloy
6	Aluminium
7	Cupro-Nickel
8	Yellow Brass
9	SPECIAL

No. Ext. Fins

Digit	
0	00
1	12
2	16
3	20
4	24
5	28
6	30
7	32
8	36
9	SPEC.
A	40
B	48

157

In calculating the U values shown in Tables 3.4 and 3.5, it was assumed that the film coefficients were similar to those used in shell-and-tube heat exchangers (Table 3.3), but these were corrected to take account of fin efficiency in the case of the fin tube units. The cost data were based on a large number of estimates for a specific application and can be expressed in terms of cost per unit area (this is total outside tube area, including the fins where appropriate) as shown in Figure 3.8.

3.4.3. Gasketed plate heat exchangers

The gasketed plate heat exchanger is illustrated schematically in the inset in Table 3.7. The gasketed plate exchanger is an extremely flexible design in that it can be custom built from standard components which can be mass produced. In the normal range, the plates are made from stainless steel, but titanium plates are commonly available and other materials can be used if necessary. The nature of the design is such that it can be optimised to give a highly efficient heat transfer performance, although this is often at the expense of a relatively high pressure drop. The gaskets are commonly made from butyl or nitrile rubber, but a variety of other compressible materials is available selected to best cope with the corrosion or temperature problems involved. These include *Neoprene*, *Viton*, *Hypalon* and ethylene-propylene.

Another major advantage of the plate exchanger is that it can be increased in size with little cost by simply adding further plates as plant requirements dictate.

Clearly, the major limitation in the plate exchanger is in its maximum operating temperature and pressure and in the integrity of sealing, which reflects the limited range of elastomer gasket materials available.

In calculating the U and C values for gasketed plate exchangers the film coefficients listed in Table 3.8 were employed. In the case of plate exchangers, the cost data can be expressed as a cost per unit area as a function of the total area of the exchanger. A curve calculated from manufacturers cost data is shown in Figure 3.9. U and C values for plate exchangers were calculated from the data in Table 3.8 and Figure 3.9 and are listed in Table 3.7 for fixed values of $Q/\Delta T$ and for the various fluid pairs.

3.4.4. Spiral heat exchangers

The principle of the spiral heat exchanger is illustrated in Figure 3.10. The hot fluid enters at the centre of the unit and flows from the inside outwards. The cold fluid enters at the periphery and flows towards the centre. The advantages claimed for this unit include the following.

(1) Since both fluids are flowing in essentially confined internal flows, the overall heat transfer coefficients are in general somewhat higher than conventional assemblies. The gaps between the plates can be adjusted to obtain the desired flow characteristics for each medium.

(2) The spiral heat exchanger, it is claimed, tends to flush away scale as it forms due to the high turbulence induced by the swirling path. This scrubbing effect permits the use of low fouling factors.

(3) The cool medium enters at the periphery and is in the outer passage so there is an effective blanket of cooler liquid surrounding the spiral assembly. Insulation is thus seldom required for the spiral exchanger.

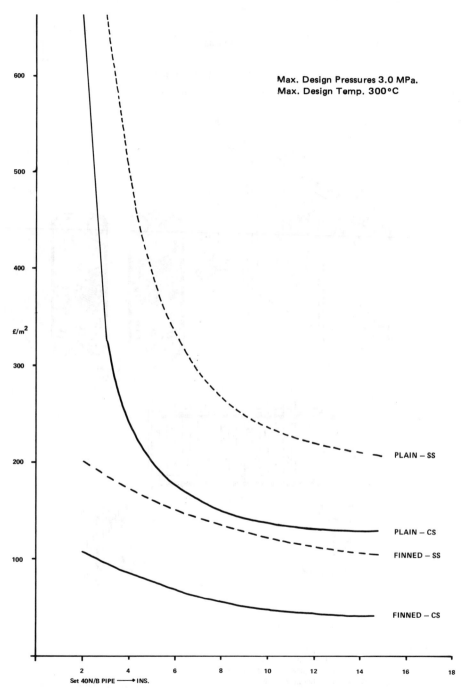

Figure 3.8—Cost curves for double-pipe heat exchangers: cost as a function of outer pipe diameter (*courtesy of Filtration and Transfer Ltd*)

TABLE 3.7

Gasketed Plate Heat Exchanger

(Courtesy of Johnson Hunt Ltd.)

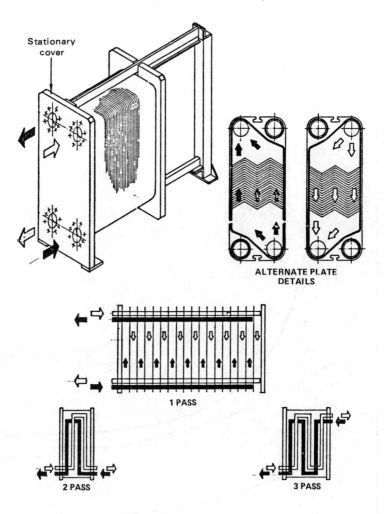

ALTERNATE PLATE
DETAILS

1 PASS

2 PASS

3 PASS

Ɨ   Normal max temp 150°C; pressure 4.4 Bar

Notes: *   Viscosity range – 5 centipoise
            **   Viscosity range   100 centipoise
            ***   Viscosity typically   1 centipoise
            NUS -   Not usually suitable
            NA -   Not applicable

			HOT SIDE FLUID							
Q/ΔT W/K	COLD SIDE FLUID	PARAMETER	Low pressure Gas -1 bar	High pressure Gas -20 bar	Process Water	Low* viscosity Organic Fluid	High viscosity Liquid	Condensing Steam	Condensing Hydrocarbon	Condensing Hydrocarbon with Inert Gas
1 000	Low pressure gas (-1 bar)	U (W/m²K)	NA	NA	NA	NA	NA	NA	NA	NA
		C (£/(W/K))	NA	NA	NA	NA	NA	NA	NA	NA
	High pressure gas (-20 bar)	U (W/m²K)	NA	NA	NA	NA	NA	NA	NA	NA
		C (£/(W/K))	NA	NA	NA	NA	NA	NA	NA	NA
	Treated Cooling Water	U (W/m²K)	NA	NA	4171	3457	331	4000	NA	NA
		C (£/(W/K))	NA	NA	0.036	0.043	0.44	0.038	NA	NA
	Low viscosity organic liquid*	U (W/m²K)	NA	NA	3163	2734	322	3063	NA	NA
		C (£/(W/K))	NA	NA	0.047	0.055	0.45	0.049	NA	NA
	High viscosity liquid**	U (W/m²K)	NA	NA	328	322	171	327	NA	NA
		C (£/(W/K))	NA	NA	0.44	0.45	0.81	0.44	NA	NA
	Boiling water	U (W/m²K)	NA	NA	NA	NA	NA	NA	NA	NA
		C (£/(W/K))	NA	NA	NA	NA	NA	NA	NA	NA
	Boiling organic liquid***	U (W/m²K)	NA	NA	NA	NA	NA	NA	NA	NA
		C (£/(W/K))	NA	NA	NA	NA	NA	NA	NA	NA
5 000	Low pressure gas (-1 bar)	U (W/m²K)	NA	NA	NA	NA	NA	NA	NA	NA
		C (£/(W/K))	NA	NA	NA	NA	NA	NA	NA	NA
	High pressure gas (-20 bar)	U (W/m²K)	NA	NA	NA	NA	NA	NA	NA	NA
		C (£/(W/K))	NA	NA	NA	NA	NA	NA	NA	NA
	Treated Cooling Water	U (W/m²K)	NA	NA	4171	3457	331	4000	NA	NA
		C (£/(W/K))	NA	NA	0.03	0.043	0.38	0.038	NA	NA
	Low viscosity organic liquid*	U (W/m²K)	NA	NA	3163	2734	322	3063	NA	NA
		C (£/(W/K))	NA	NA	0.047	0.055	0.40	0.045	NA	NA
	High viscosity liquid**	U (W/m²K)	NA	NA	328	322	171	327	NA	NA
		C (£/(W/K))	NA	NA	0.39	0.40	0.68	0.39	NA	NA
	Boiling water	U (W/m²K)	NA	NA	NA	NA	NA	NA	NA	NA
		C (£/(W/K))	NA	NA	NA	NA	NA	NA	NA	NA
	Boiling organic liquid***	U (W/m²K)	NA	NA	NA	NA	NA	NA	NA	NA
		C (£/(W/K))	NA	NA	NA	NA	NA	NA	NA	NA
30 000	Low pressure gas (-1 bar)	U (W/m²K)	NA	NA	NA	NA	NA	NA	NA	NA
		C (£/(W/K))	NA	NA	NA	NA	NA	NA	NA	NA
	High pressure gas (-20 bar)	U (W/m²K)	NA	NA	NA	NA	NA	NA	NA	NA
		C (£/(W/K))	NA	NA	NA	NA	NA	NA	NA	NA
	Treated Cooling Water	U (W/m²K)	NA	NA	4171	3457	331	4000	NA	NA
		C (£/(W/K))	NA	NA	0.033	0.039	0.26	0.034	NA	NA
	Low viscosity organic liquid*	U (W/m²K)	NA	NA	3163	2734	322	3063	NA	NA
		C (£/(W/K))	NA	NA	0.042	0.049	0.26	0.044	NA	NA
	High viscosity liquid**	U (W/m²K)	NA	NA	328	322	171	327	NA	NA
		C (£/(W/K))	NA	NA	0.26	0.26	0.43	0.26	NA	NA
	Boiling water	U (W/m²K)	NA	NA	NA	NA	NA	NA	NA	NA
		C (£/(W/K))	NA	NA	NA	NA	NA	NA	NA	NA
	Boiling organic liquid***	U (W/m²K)	NA	NA	NA	NA	NA	NA	NA	NA
		C (£/(W/K))	NA	NA	NA	NA	NA	NA	NA	NA
100 000	Low pressure gas (-1 bar)	U (W/m²K)	NA	NA	NA	NA	NA	NA	NA	NA
		C (£/(W/K))	NA	NA	NA	NA	NA	NA	NA	NA
	High pressure gas (-20 bar)	U (W/m²K)	NA	NA	NA	NA	NA	NA	NA	NA
		C (£/(W/K))	NA	NA	NA	NA	NA	NA	NA	NA
	Treated Cooling Water	U (W/m²K)	NA	NA	4171	3457	331	4000	NA	NA
		C (£/(W/K))	NA	NA	0.029	0.038	0.19	0.030	NA	NA
	Low viscosity organic liquid*	U (W/m²K)	NA	NA	3163	2734	322	3063	NA	NA
		C (£/(W/K))	NA	NA	0.037	0.041	0.19	0.038	NA	NA
	High viscosity liquid**	U (W/m²K)	NA	NA	328	322	171	327	NA	NA
		C (£/(W/K))	NA	NA	0.19	0.19	0.32	0.19	NA	NA
	Boiling water	U (W/m²K)	NA	NA	NA	NA	NA	NA	NA	NA
		C (£/(W/K))	NA	NA	NA	NA	NA	NA	NA	NA
	Boiling organic liquid***	U (W/m²K)	NA	NA	NA	NA	NA	NA	NA	NA
		C (£/(W/K))	NA	NA	NA	NA	NA	NA	NA	NA
1000 000	Low pressure gas (-1 bar)	U (W/m²K)	NA	NA	NA	NA	NA	NA	NA	NA
		C (£/(W/K))	NA	NA	NA	NA	NA	NA	NA	NA
	High pressure gas (-20 bar)	U (W/m²K)	NA	NA	NA	NA	NA	NA	NA	NA
		C (£/(W/K))	NA	NA	NA	NA	NA	NA	NA	NA
	Treated Cooling Water	U (W/m²K)	NA	NA	4171	3457	331	4000	NA	NA
		C (£/(W/K))	NA	NA	0.015	0.018	0.14	0.016	NA	NA
	Low viscosity organic liquid*	U (W/m²K)	NA	NA	3163	2734	322	3063	NA	NA
		C (£/(W/K))	NA	NA	0.020	0.022	0.14	0.020	NA	NA
	High viscosity liquid**	U (W/m²K)	NA	NA	328	322	171	327	NA	NA
		C (£/(W/K))	NA	NA	0.14	0.14	0.26	0.14	NA	NA
	Boiling water	U (W/m²K)	NA	NA	NA	NA	NA	NA	NA	NA
		C (£/(W/K))	NA	NA	NA	NA	NA	NA	NA	NA
	Boiling organic liquid***	U (W/m²K)	NA	NA	NA	NA	NA	NA	NA	NA
		C (£/(W/K))	NA	NA	NA	NA	NA	NA	NA	NA

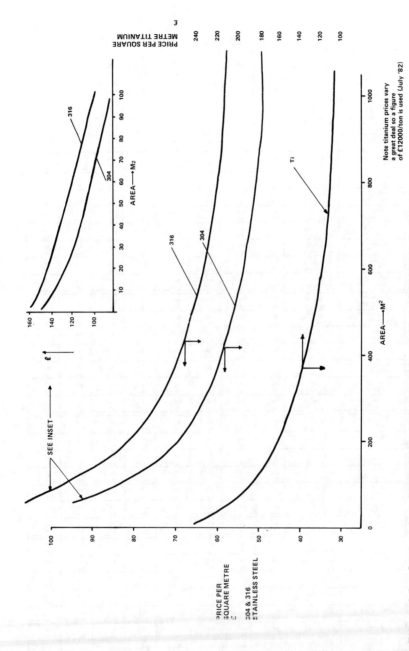

Figure 3.9—Cost curves for gasketed plate heat exchangers constructed from stainless steel types 304 and 316 and from titanium (*courtesy of Johnson Hunt Ltd*)

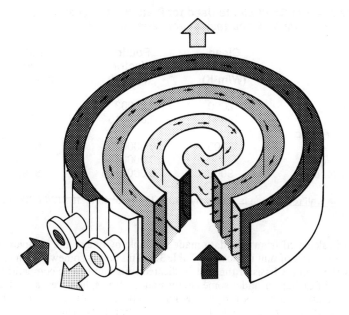

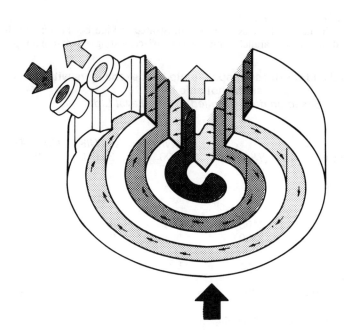

Figure 3.10—Flow paths in a spiral heat exchanger (*courtesy of Alfa-Laval Ltd*)

Table 3.8—Film Coefficients Used for Plate Heat Exchangers
Basic data by courtesy of Johnson Hunt Ltd.

	Clean coefficient (W/m²K)	Fouling factor (Km²/W)	Overall film coefficient (W/m²K)
Process water	12000	0.00003	8824
Treated water	14000	0.000015	11570
Low viscosity organic liquids	7000	0.00002	6140
High viscosity liquid	350	0.00004	345
Steam (low pressure)	9000	0.0000125	8090

Note also that the wall resistance is relatively high a figure of 40×10^{-6} (W/m²K)$^{-1}$ for stainless is used.

U and C values calculated from cost data made available by one of the companies manufacturing this form of unit is shown in Table 3.9. This data is somewhat limited but does show that, on the basis of the costs indicated, this unit can be competitive with other forms of exchanger under some circumstances. It has, however, a limited range of pressure and within this range, it is not competitive with gasketed plate exchangers where the latter are applicable. However, it may find a role in an intermediate temperature range and for situations where a suitable gasket material is not available.

3.4.5. Lamella heat exchangers
The principle of the lamella heat exchanger is illustrated in Figure 3.11. One of the fluids passes inside a series of parallel "lamella elements" consisting of two thin dimpled

Table 3.9—U and C values for spiral plate heat exchangers with water as the hot side fluid
(Based on data supplied by Alfa-Laval Ltd.)

Cold side fluid	U (W/m²K)	C $Q/\Delta T$ (W/K) = 1000	$Q/\Delta T$ (W/K) = 30 000	$Q/\Delta T$ (W/K) = 100 000
Low pressure gas (~ 1 bar)	240	*	1.25	*
High pressure gas (~ 20 bar)	1100	*	0.3	0.27
Water	2500	*	0.13	0.13
Low viscosity organic	1000	*	0.33	0.3
High viscosity liquid	85	3.88	*	*
Boiling water	2700	*	0.12	0.12
Boiling organic liquid	1700	*	0.19	0.19

*Outside range of size or available data.

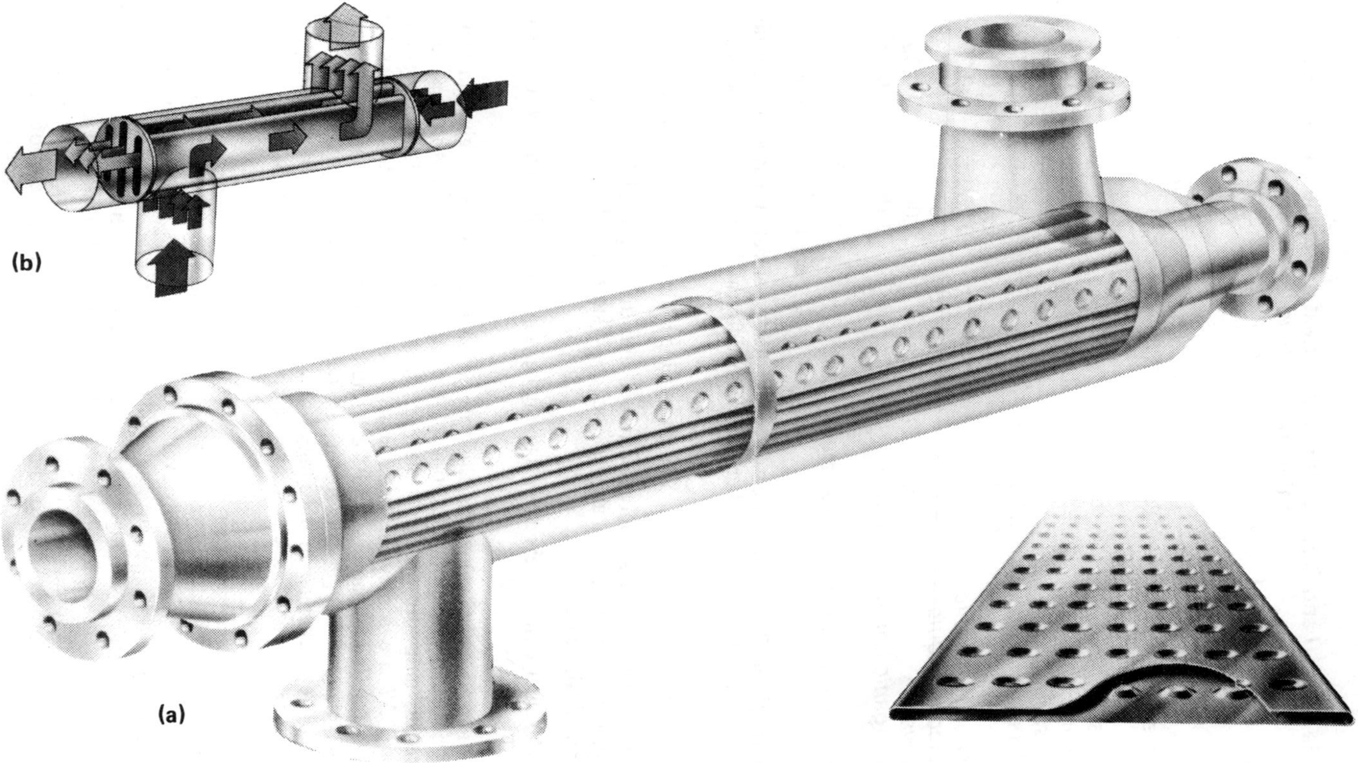

Figure 3.11—The lamella heat exchanger (*courtesy of Alfa-Laval Ltd*)

Table 3.10—*U* and *C* values for lamella heat exchangers with water as the hot side fluid

(Based on data supplied by Alfa-Laval Ltd.)

Cold side fluid	U (W/m²K)	C		
		$Q/\Delta T$ (W/K) = 1000	$Q/\Delta T$ (W/K) = 30 000	$Q/\Delta T$ (W/K) = 100 000
Low pressure gas (~ 1 bar)	275	1.05	0.78	0.78
High pressure gas (~ 20 bar)	1300	*	0.22	0.22
Water	3000	*	0.09	0.09
Low viscosity organic	1150	*	0.25	0.25
High viscosity liquid	100	2.90	1.87	*
Boiling water	3250	*	0.09	0.09
Boiling organic liquid	1950	*	0.15	0.15

*Outside range of size or available data

metal strips (Figure 3.11c) which are welded together in pairs at the edges, thus forming a straight channel or lamella. The benefits claimed for this heat exchanger include the following:

(1) The exchanger is in pure counter-flow giving a higher thermal efficiency.

(2) The uniform flow distribution achieved and the long straight flow channels reduce pressure drops and allow suspensions, slurries and fibrous liquids to be handled with ease.

(3) The exchanger has a very compact design giving small units with low volumetric holdup.

Approximate U and C values for this form of heat exchanger are given in Table 3.10 based on data supplied by one of the manufacturers. It is interesting to compare this unit with the spiral plate exchanger (see Table 3.9). The costs for a given $Q/\Delta T$ are somewhat lower but it may be appropriate to select this exchanger for its compactness and for its ability to handle suspensions. As with the spiral plate exchanger, this form of exchanger is not competitive with the gasketed plate exchanger in the range where the latter can be employed. However, it can operate at somewhat higher temperatures (see Table 3.1).

3.4.6. *Welded plate exchanger*
The welded plate exchanger overcomes the gasket problem encountered with gasketed plate exchangers. A commercially available welded plate exchanger construction is illustrated in Figure 3.12. For such units, higher pressures (equivalent to those applicable in shell-and-tube exchangers) can be achieved by mounting the plates within a shell. Very large areas are feasible (> 1000m²) and relatively high fluid temperatures can be coped with. Typical plates in such units can be up to 10 m long and 1.5 m wide; and the large size compensates to some extent for the increased cost of sealing.

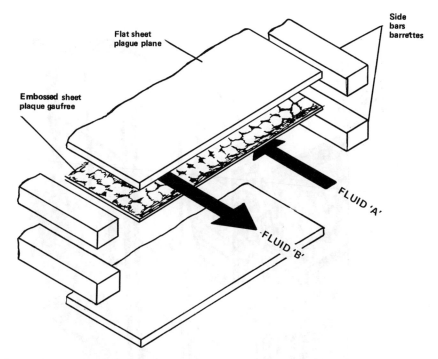

Figure 3.12—Welded-plate heat exchanger construction (*Packinox* heat exchanger) (*courtesy of Nouvelles Applications Technologiques, France*)

Costs for this form of exchanger are obviously higher than those of the gasketed plate exchanger, but by using large plate sizes, some manufacturers claim that the increase is no more than, say 10 to 20%.

A limitation on this form of exchanger is the differential pressure between the two fluids; a typical figure given by a manufacturer for maximum differential pressure is 3.0MPa.

An often overriding problem with this form of exchanger is that chemical rather than mechanical cleaning of the plates is mandatory. This rules out many applications.

3.4.7. Hot gas-to-liquid convective bank systems
In the recovery of waste heat from the process streams, a very wide variety of units is available in which the waste heat is transferred to a liquid phase, often water. Typical of the types of units employed are those illustrated in Figures 3.13 to 3.15.

Type 1 (Figure 3.13)—This consists of a matrix of tubes with widely spaced rectangular fins either of cast iron or welded carbon steel. Such units may operate up to temperatures of around 700°C, though oxidation becomes significant over 600°C.

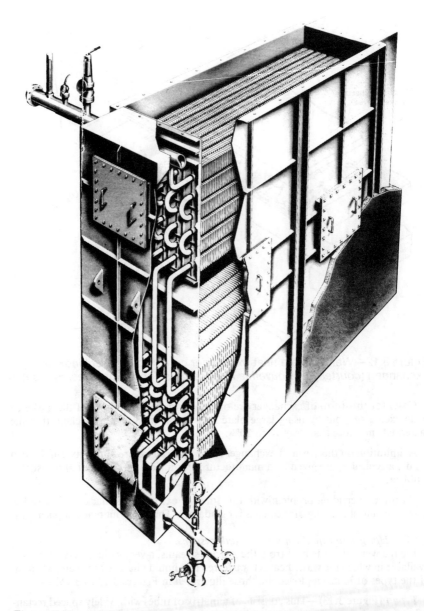

Figure 3.13— High-temperature-gas convective bank exchanger with finned tubes (gas-to-liquid exchanger Type 1) (*courtesy of E. Green and Son Ltd*)

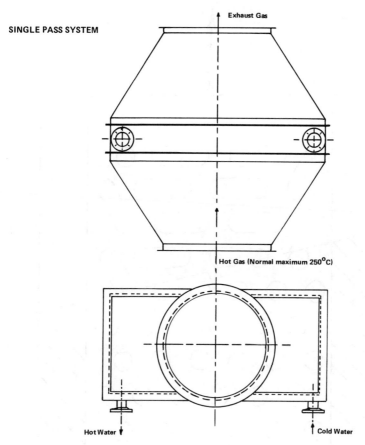

SINGLE PASS SYSTEM

Exhaust Gas

Hot Gas (Normal maximum 250°C)

Hot Water

Cold Water

MULTI PASS SYSTEM

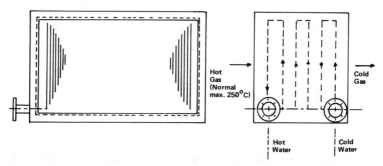

Hot Gas (Normal max. 250°C)

Cold Gas

Hot Water

Cold Water

Figure 3.14—Extended-surface cross-flow exchanger (gas-to-liquid exchanger type 2) (*courtesy of Hunt Heat Exchangers Ltd*)

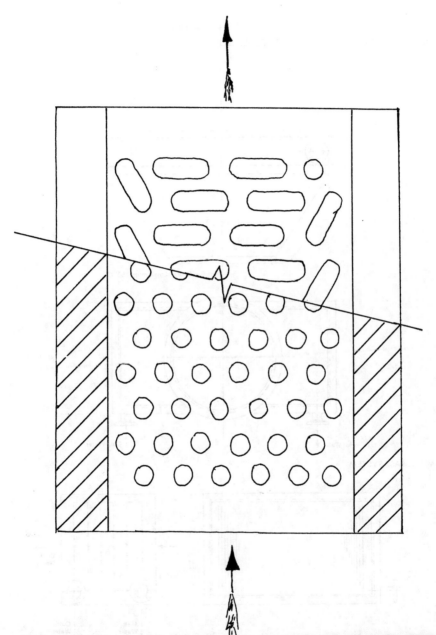

Figure 3.15—High-temperature-gas convective bank exchanger with plain tubes with ceramic-lined containment constructed on site (gas-to-liquid exchanger type 3; gas-to-gas exchanger type 2)

Type 2 (Figure 3.14)—Here the gas passes over finned tubes mounted in a metal construction as shown, operating either with single or multiple passes on the liquid side. This type of unit operates up to temperatures of the order of 250°C.

Type 3 (Figure 3.15)—Plain tube unit in site-built refractory lined casing. This type of unit can operate up to very high temperatures, particularly when refractory metals are used for the tubes.

When using these types of unit for heat recovery from combustion gases, consideration has to be given to the corrosion which may take place if the external surface of the tubes is below the dew point of the gases. It is usual practice to ensure that the cold fluids enter the convection bank at a temperature in excess of that at which the condensation of acidic moisture will occur.

Approximate U and C values for the various types of gas-to-liquid convective bank systems are given in Table 3.11. It will be seen from this that the light construction of Type 2 makes it a very much cheaper option than the other types. Type 3 (with its on-

Table 3.11—U and C values for hot gas-to-liquid convective bank systems (water or other low viscosity liquid on the tube side)

Type 1 (Figure 3.13) with carbon steel tubes and cast iron fins	$Q/\Delta T$ (W/K)	2000	12 000	50 000
	U (W/m^{2}K)	25	25	26
	C £/(W/K)	7.4	3.7	2.5
Type 1 (Figure 3.13) with carbon steel tubes and carbon steel welded fins	$Q/\Delta T$ (W/K)	5000	25 000	—
	U (W/m^{2}K)	20	24	—
	C £/(W/K)	4.7	2.0	—
Type 2 (Figure 3.14) (with extended surface tubes)	$Q/\Delta T$ (W/K)	15 000	7000	—
	U (W/m^{2}K)	85	85	—
	C £/(W/K)	0.20	0.13	—
Type 3 (Figure 3.15) cast iron tubes	$Q/\Delta T$ (W/K)	1000	1000	—
	U (W/m^{2}K)	17	17	—
	C £/(W/K)	19	16	—
Type 3 (Figure 3.15) carbon steel tubes	$Q/\Delta T$ (W/K)	1000	20 000	—
	U (W/m^{2}K)	17	17	—
	C £/(W/K)	17	14	—
Type 3 (Figure 3.15) stainless steel tubes	$Q/\Delta T$ (W/K)	1000	20 000	—
	U (W/m^{2}K)	17	17	—
	C £/(W/K)	25	22	—

site construction in a ceramic liner) would only be used when temperature or other process conditions dictate it.

It should be remembered that installation costs (including ducting and dampers) can vary widely, especially for high temperature gases, and can appreciably affect the total cost of the system.

3.4.8. Gas-to-gas recuperative heat exchangers

Gas-to-gas recuperative heat exchangers find very wide use in commercial and industrial applications. Perhaps the most widely used form is that which uses combustion gases to preheat furnace air. Other applications include heat recovery from building exhaust air, feed-effluent heat exchangers for gaseous chemical reactors *etc*. A wide variety of types of unit are available; it will suffice here to list and discuss some of the principal types.

Type 1 (Figure 3.16)—Plain tube air-to-air heat exchanger. As will be seen from Figure 3.16, this is of usually light construction with a plain tube bank, the air passing through the tubes and the hot gas passing over them. A normal maximum operation temperature for such units is 250°C. The units can be of the single-pass (cross flow) type or of the multi-pass type as illustrated. For the single-pass type, a major disadvantage is that the cross flow correction factor (F_T) is low and this leads to low effective temperature differences (see equation 5). For the multi-pass system, F_T is much higher. A typical U-value for such a system would be 35 W/m²K and for large units, C is around 1.4 £/(W/K).

Type 2 (Figure 3.15)—High-temperature gas convective banks with plain tubes. This form of unit is constructed in site-built refractory casings and can be used for gas-to-gas heat transfer as well as for gas-to-liquid heat transfer as mentioned above. A typical value of the overall heat transfer coefficient for gas-to-gas heat transfer in such a unit would be around 14 W/m²K. For cast iron tubes, C would be of the order of 21 £/(W/K), for carbon steel tubes around 17 £/(W/K) and for stainless steel tubes around 28 £/(W/K). As will be seen, the cost of these units is rather high compared to off-site built units, but such units have to be used under certain circumstances. The maximum temperature of operation for this type depends on the materials of construction. With high-temperature alloys, it could be in excess of 1000°C. Again, since the unit is a cross flow unit, F_T is low leading to a much lower effective temperature driving force.

Type 3 (Figure 3.17)—Hot gas-to-air recuperator with cast iron elements. This is a traditional form of design for a recuperator and is constructed from modular elements as illustrated. These units tend to be very large and range typically in duty from 12 000 W/K to 300 000 W/K. A typical overall heat transfer coefficient (based on the finned area) would be of the order of 11 W/m²K and C ranges typically from 3.5 £/(W/K) for the smaller units down to 2.0 £/(W/K) for the largest duties. Typical operating temperatures range up to 400°C. These units are, therefore, considerably cheaper than Type 2 units since they can be built in a modular form off-site. The use of a finned surface gives a greater propensity to fouling than does the use of plain tubes.

Type 4—Glass tube heat exchangers. These exchangers are suitable for recovery of heat from aggressive exhaust gases at temperatures up to around 250°C. The glass

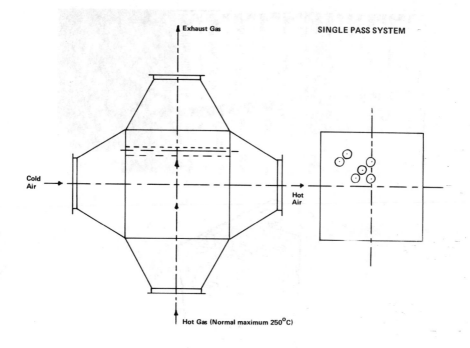

SINGLE PASS SYSTEM

Exhaust Gas

Cold Air

Hot Air

Hot Gas (Normal maximum 250°C)

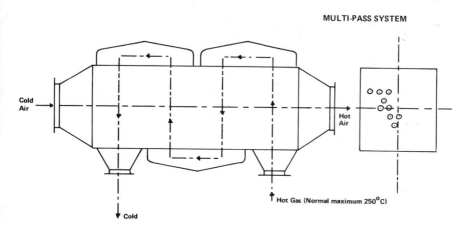

MULTI-PASS SYSTEM

Cold Air

Hot Air

Hot Gas (Normal maximum 250°C)

Cold

Figure 3.16—Plain tube air-to-air heat exchanger (gas-to-gas exchanger type 1) (*courtesy of Hunt Heat Exchangers Ltd*)

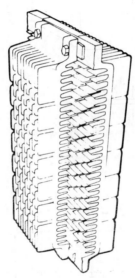

Figure 3.17—Hot-gas-to-air recuperator with cast-iron elements (gas-to-gas exchanger type 3) *(courtesy of E. Green and Son Castings Ltd)*

tubes are mounted horizontally between perforated stainless steel end-plates into which they are securely sealed. Plastic spacers hold the tubes in correct location. The end-plates, framework and flange enclosure are fabricated from chemical-resistant stainless steel.

Type 5—Convection banks with plastic tubes. Here PVC-C tubes are mounted horizontally between perforated stainless steel end-plates into which they are securely sealed. Plastic spacers hold the tubes in correct location and the end-plates, framework and flanged enclosure are fabricated from high-grade chemical-resistant stainless steel. Other plastic tubes (*e.g.* PTFE) are available to special order. These units are suitable for recovery of heat from aggressive exhaust gases at temperatures up to 100°C.

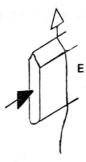

Figure 3.18—Flat stainless steel tube element used in gas-to-gas convection banks, type 6

Type 6—Convection banks with flat stainless steel tubes. In a typical unit, flat-sided tubes (see Figure 3.18) are mounted horizontally and silicon-sealed into perforated end-plates formed into a flange plate framing and enclosure. The material throughout is stainless steel. The tubes are formed from polished, dimpled-ribbed stainless steel and spaced to give clear wide vertical air passages which encourage any condensate to wash out all deposits clear from the base of the unit. The units are very compact and light weight and are suitable for pressure differentials up to 1200 Pa. They can be used for the recovery of heat from exhaust gases up to 250°C and are especially effective in self-cleaning of particulate solids contained in high humidity exhaust air. They have applications in the chemical industry, paper manufacturing, the dairy industry, textiles, malt kilns and similar batch drying processes.

Type 7—Gas convection banks of the plate type. The principle of the plate type heat exchanger is illustrated in the sketch in Figure 3.19. The hot gas and the air are passed through alternate gaps between plates as illustrated. Obviously, such a plate-type unit can be constructed from a wide variety of materials. Some cost data were obtained in the present exercise for a unit constructed from special aluminium plates with pressed metal spacing lugs. The plates are metal-sealed into exchanger elements which are assembled and silicone or metal sealed into a flanged aluminium plate housing. The exchanger element plates are widely spaced with straight-through passages for ease of cleaning and the units are pressure-tight up to 6000 Pa. For general air-to-air applications, untreated aluminium plates can be employed but with mildly aggressive hot side fluids, the aluminium can be epoxy coated.

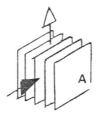

Data courtesy of
Food and Beverage
Development (UK) Ltd.

Figure 3.19—Plate heat exchanger form used in gas-to-gas convection banks, type 7

Table 3.12—C-values for gas-to-gas recuperative heat exchangers

	C-values for		
	$Q/\Delta T$ (W/K) = 1000	$Q/\Delta T$ (W/K) = 4000	$Q/\Delta T$ (W/K) = 30 000
Type 4 (glass tube) with low pressure loss (50–125 Pa)	1.94	1.97	1.40
Type 4 (glass tube) with high pressure loss (250–400 Pa)	0.78	0.72	0.53
Type 5 (plastic tube) with low pressure loss (50–125 Pa)	1.56	1.29	1.02
Type 5 (plastic tube) with high pressure loss (250–400 Pa)	0.85	0.45	0.36
Type 6 (flat stainless tubes) with low pressure loss (50–125 Pa)	4.60	1.53	0.53
Type 6 (flat stainless tubes) with high pressure loss (250–400 Pa)	—	1.15	0.36
Type 7 (aluminium plate) with untreated plates	0.43	0.28	0.23
Type 7 (aluminium plate) with epoxy-coated plates	0.50	0.32	0.26

Data were obtained from which C-values could be calculated for Type 4 to Type 7 units. These are presented in Table 3.12. According to these data, the aluminium plate units are the cheapest of this range of types, even when coated with epoxy resin. However, the corrosive nature of the hot gas stream may dictate the use of one of the other types. It should be noted that all of the cross flow units have a relatively low value of F_T giving a reduced effective temperature driving force.

3.4.9. Heat-pipe heat exchangers

The heat-pipe heat exchanger is illustrated in Figure 3.20. Heat is transferred from the hot gas to the evaporation section. The vapour flows along the heat pipe to the condensation section, releasing its heat to the cold gas. The liquid returns along the wick to the evaporation section. A series of heat pipes can be mounted as shown in Figure 3.20(b) and, in this form, the heat-pipe heat exchanger operates in a mode close to counter-current heat exchange.

The heat-pipe heat exchangers are normally operated with the pipes in the vertical position (condensation section at the top) or with a slight inclination from the horizontal.

Most heat pipe systems are operated with water as the heat pipe fluid. Higher temperatures are possible with organic fluids, though thermal degradation of these can be a problem. Very high temperatures (up to 1500°C) are feasible if a liquid metal such as sodium is used as the fluid but this is often too high a range.

More fluids suitable for the intermediate temperature range are currently under development.

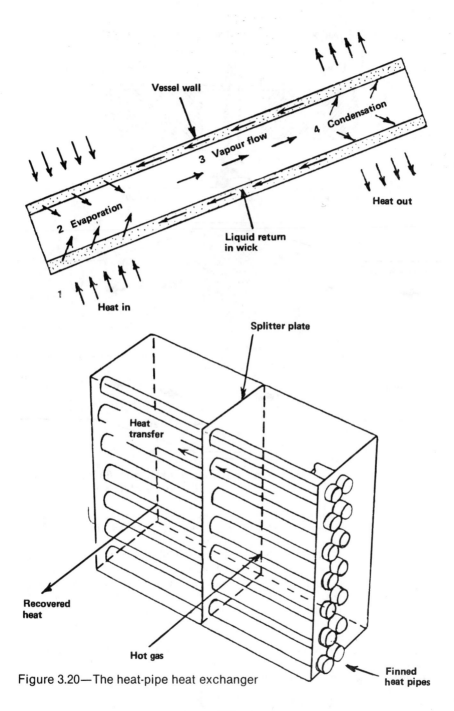

Figure 3.20—The heat-pipe heat exchanger

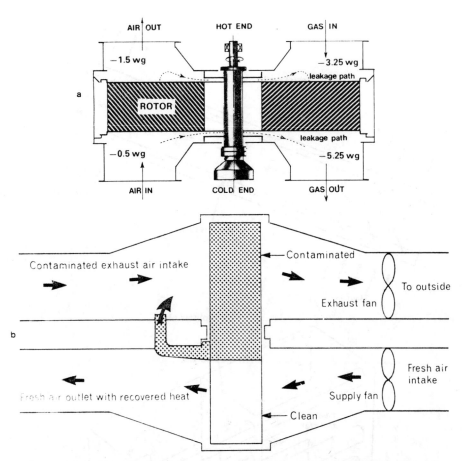

Figure 3.21—Forms of rotary regenerative heat exchanger ((a) *courtesy of James Howden and Co Ltd;* (b) *courtesy of Curwen and Newbery Ltd*)

The overall heat transfer coefficients for units of this type are of the order of 20 W/m²K. Duties (expressed as $Q/\Delta T$) typically range from 2000 to 20 000 W/K. For copper heat pipes, the C values are typically 0.5 £/(W/K) for the smaller sizes going down to 0.2 £/(W/K) for the bigger size units. With carbon steel heat pipes, the equivalent range of C values is 1.0 to 0.4 £/(W/K) for the same range of duties. These figures are, of course, approximate; but they do give an indication that the heat-pipe heat exchanger is already quite competitive for many applications. The main reasons for this promising performance are that both of the gas sides can employ extended surfaces. Furthermore, the heat-pipe heat exchanger sometimes lends itself to easier installation within processes from the point of view of installation within duct-work. If the hot gas duct may be placed below the cold gas duct, vertical heat pipes may be used; here, a wickless heat pipe ("thermosyphon") can be employed. Otherwise, the heat pipes should be nearly horizontal with the evaporator lower than the condenser.

3.4.10. Rotary regenerators

In this form of apparatus heat from a hot gas stream is imparted to a metallic or ceramic medium on a slowly rotating wheel. The rotation of the wheel brings it into contact with a cold gas stream, to which the medium loses the heat which it has collected. This type of equipment can be substantially more compact and less heavy than the equivalent recuperative heat exchanger for the duty.

The direct method of heat transfer results in uniformally higher temperatures in contact with the hot gases than in the recuperative type of exchanger. Thus, lower hot gas stream exit temperatures can be achieved with minimised danger of acidic attack, an important point in favour of this form of waste heat recovery system.

A disadvantage of rotary regenerators is cross contamination between the gas streams. However, this can be greatly mitigated by means of devices such as double labyrinth seals at the periphery and by utilising a purge unit in conjunction with air seals.

The electrical power consumed in driving the rotor is usually negligible in comparison with the thermal energy recovered. Typical forms of rotary regenerators are illustrated in Figure 3.21. By the courtesy of the manufacturers of these units, approximate cost data have been obtained. These can be expressed in terms of the C value for a given duty, expressed in terms of the ratio $Q/\Delta T$. For the form of units shown in Figure 3.21a, the data are given in Table 3.13; for the form shown in Figure 3.21b, the cost data are shown in Table 3.14. It will be seen from these tables that the cost of regenerative heat exchangers are generally higher than those of recuperative heat exchangers for the same duties and operating temperature range. However, regenerative heat exchangers are still often chosen by reason of their compactness and because of the high surface temperatures, and avoidance of acidic corrosion, as mentioned above. For high operating temperatures, the regenerative heat exchanger can be competitive notwithstanding these factors, due to the high installation costs of special recuperative exchangers (see Section 3.4.7 and 3.4.8 above).

It should be stressed that the values given in Tables 3.13 and 3.14 are for ex-works costs; the installation costs (including ducting and dampers) can vary widely (especially for high temperature gases) and can appreciably affect the total cost of the system.

Table 3.13—Typical cost data for rotary regenerative heat exchangers of the form shown in Figure 3.21a
(Basic data by courtesy of James Howden and Co Ltd)

Typical Gas Temperature (°C)		Duty area range (W/K)	Electrical power required (KW)	C values £/(W/K)
Hot gas	Cold gas			
520 → 300	20 → 276	3 179	0.75	3.775
454 → 249	46 → 283	20 806	1.5	2 312
335 → 124	32 → 291	1.225×10^6	15.0	1.173

Table 3.14—Typical *C*-value data for rotary regenerative heat exchangers of the form shown in Figure 3.21b
(Basic data by courtesy of Curwen and Newbery Ltd.)

Materials		Typical hot gas temperatures (°C)	Duty range (W/K)	Electrical power required (KW)	C value £/(W/K)
Rotor	Transfer media				
Aluminium	Aluminium	< 200	1000	0.18	1.857
			30 000	1.49	0.80
			1×10^6	33.6	0.714
Mild Steel	316 Stainless Steel	< 427	1000	0.18	2.686
			30 000	1.49	1.171
			1×10^6	33.6	1.029
316 Stainless Steel	316 Stainless Steel	< 427	1000	0.18	3.20
			30 000	1.49	1.40
			1×10^6	33.6	1.257
310 Stainless Steel	310 Stainless Steel	< 980	1000	0.18	7.114
			30 000	1.49	3.114
			1×10^6	33.6	2.743

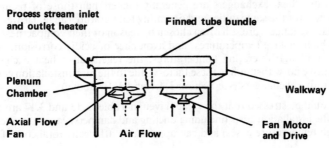

(a) Forced draft unit

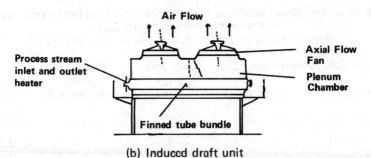

(b) Induced draft unit

Figure 3.22—Air-cooled heat exchangers

3.5. Localised Utilities

3.5.1. Air cooled heat exchangers

Although air cooled heat exchangers are essentially a *heat rejection* rather than a *heat exchange* device, they have great importance in process plant and information on their cost is important in considering the overall thermal integration of such plant. Note that air cooled heat exchangers are normally primarily cross flow units and hence have a relatively low value of F_T. The penalty for cross flow can be minimised by using multiple tube-side passes.

Typical air cooled heat exchanger units are illustrated in Figure 3.22 for forced-draught and induced-draught operation respectively. The induced-draught mode of operation tends to give a uniform flow distribution, but the circulation fans have to be designed to operate in a warm air stream.

Typical air cooled heat exchanger tubing would be 1 inch (25.4 mm) outside diameter carbon steel tube wound with aluminium fins, at 11 fins per inch with a 2 inch (50.8 mm) outside diameter. For such units, the bundle costs represent around 60% of the total unit cost. As with other forms of exchanger, the cost decreases with increasing surface area and data for air cooled heat exchangers are illustrated in Figure 3.23 where the

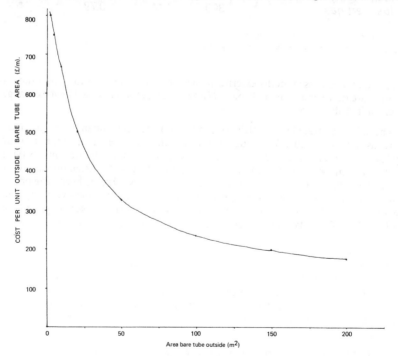

Figure 3.23—Typical cost data for air-cooled heat exchangers with 1-inch-diameter carbon-steel tubes wound with 11 2-inch-diameter aluminium fins per inch (*courtesy of Lummus (UK) Ltd*)

Table 3.15—U and C values for air cooled heat exchangers with standard 1 inch diameter carbon steel tubes with 2 inch diameter aluminium fins at 11 fins per inch
(Data by courtesy of Lummus (UK) Ltd.)

Process side fluid	U^* (W/m²K)	$Q/\Delta T$ = 1000	C (£/(W/K)) $Q/\Delta T$ = 30 000	$Q/\Delta T$ = 100 000
Low pressure gas (~ 1 bar)	85	0.65	**	**
High pressure gas (~ 20 bar)	450	**	0.65	**
Treated cooling water	700	**	0.51	0.31
Low viscosity organic liquid	550	**	0.56	0.33
High viscosity liquid	100	6.7	**	**
Condensing steam	800	**	0.48	0.26
Condensing hydrocarbon	500	**	0.60	0.35
Condensing hydrocarbon + H₂	450	**	0.64	**
Condensing hydrocarbon plus inert gas	300	**	0.79	**

* Referred to outside of bare tube
** Beyond range of data available (5–200 m²)

cost per unit area (based on the external area of the bare tubes) is shown as a function of total bare tube surface area. Typical U and C values for air cooled heat exchangers are given in Table 3.15.

Tube-side heat transfer coefficients can be considerably enhanced by the use of inserts as shown in Figure 3.24. Such inserts also increase pressure drop, but in some applications this effect can be minimised by using different configurations. A typical cost for such inserts would be £1/m and the cost of the insert is often only a few per cent of the total cost of the heat exchanger. Inserts of this form are becoming popular in air cooled heat exchangers for high viscosity liquids. For the high viscosity liquid case shown in Table 3.15, the C value would be reduced, using these inserts, from 6.7 to 4.0 for $Q/\Delta T = 1000$. Such inserts could also be used in other tubular exchangers.

3.5.2. Cooling towers
As an alternative to site-supplied cooling water, small localised units may be employed of the form shown in Figure 3.25. Air is drawn through the unit by the fan (1) over the packing (6) contained in a corrosion-resistant shell (4). The cooling water enters through the spray nozzle (2) and flows over the packing counter-current to the air and collects at the bottom of the unit (5) from where it passes back to the process heat exchanger.

Cooling towers involve complex processes of heat and mass transfer but they can be treated tentatively in a manner analogous to the treatment given above for heat exchangers. Typically, cooling towers are designed for wet bulb air inlet and outlet temperatures of 17°C and 21°C respectively. The process can be considered very

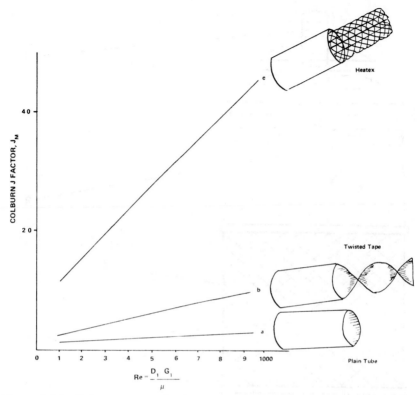

Figure 3.24—Enhancement of tube-side heat transfer coefficients using inserts (*courtesy of Cal Calvin Ltd*)

approximately as one of counter-current heat transfer between the air and the cooling water, though, of course, around 80% of the heat transfer is in the form of evaporation of the cooling water. Using the 17°C to 21°C wet bulb air temperature range, values of C were calculated from data supplied by a manufacturer and are given in Table 3.16. It should be stressed strongly that these figures are given for order-of-magnitude guidance only.

In considering the overall costs of cooling towers compared to other units, account should be taken of two factors:

(1) The cost of fan power.

(2) The cost of the cooling water lost by evaporation. This can be significant in some geographical locations.

An example comparison between the use of a shell-and-tube heat exchanger coupled with a cooling tower and the alternative direct use of an air-cooled heat exchanger is given in Section 3.6.4.

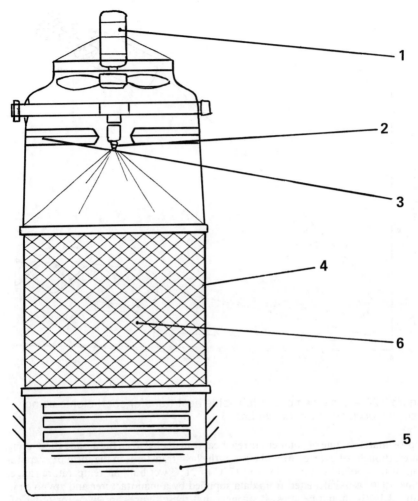

Figure 3.25—Typical packaged cooling tower (*courtesy of Davenport Engineering Co Ltd*)

Table 3.16—*C* values for small forced-draught cooling towers
(Basic data by courtesy of Davenport Engineering Co. Ltd.)

$Q/\Delta T$ (W/K)	C (£/(W/K))
$Q/\Delta t = 1000$	0.7
$Q/\Delta t = 30\ 000$	0.048
$Q/\Delta t = 1\ 000\ 000$	0.029

3.5.3. Package boilers

A wide variety of package boilers is available and no attempt will be made here to give indicative costs since these vary so widely between boiler types, with boiler pressure and with the type of fuel being employed. Obviously, waste heat streams can often be used to generate steam. Waste heat boilers are available in a variety of types and their application has given rise to a number of problems including tube sheet failure due to local overheating and gas side fouling. An extensive review of waste heat boiler types and problems is given by Hinchley (1977).

Depending on the power requirements of a particular plant, and of adjacent plants, it may well be worthwhile considering combined heat and power generation (CHP).

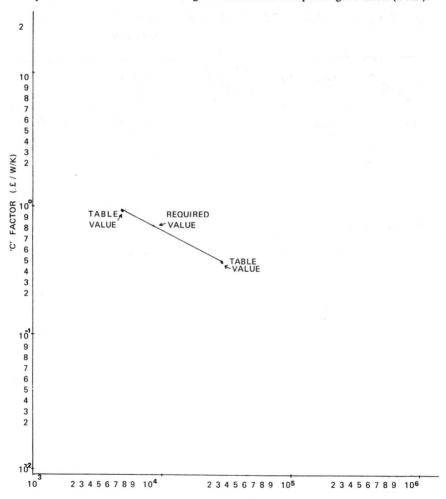

Figure 3.26—Interpolation graph for C-values

3.6. Examples of Application

3.6.1. Oil cooler
Example 1—A heat exchanger is required for the cooling of a high viscosity oil with a specific heat of 2 kJ/kgK flowing at a rate of 2.5 kg/s and at a pressure of 3.0 MPa. The oil enters the exchanger at 70°C and leaves at 30°C. It is cooled by water flowing at 10 kg/s entering at 20°C and leaving at 24.8°C.

In this case the pressure is higher than would be permissible for a gasketed plate heat exchanger. The choice is therefore between a shell and shell-and-tube heat exchanger and a double-pipe heat exchanger.

$$Q = W_1 C_{p1} (T_{1i} - T_{1o}) = 2.5 \times 2 (70 - 30) = 200 \text{ kJ/s}$$
$$= 200\ 000 \text{ Watt}$$

$$\Delta T = \frac{45.2 - 10}{\log_e \dfrac{45.2}{10}} = 23.33 \text{ °K}$$

70

(40)

30 (45.2)

(10) 24.8

(4.8)

20

(50)

$$R = \frac{40}{4.8} = 8.33; \quad P = \frac{4.8}{50} = 0.096$$

$$\therefore F_T \text{ (from Figure 3.2)} = 0.93$$

Shell and Tube	Double Pipe

ΔT for 2 pass tubeside = 21.7

ΔT (true counterflow) = 23.33

$Q/\Delta T$ = 9217 W/K

$Q/\Delta T$ = 8572

	Shell and Tube	Double Pipe Finned	Double Pipe Plain
from Table 3.2	$U = 142$	$U = 108$	142
$C_{(Q/\Delta T = 5000)}$	0.94	.46	1.232
$C_{(Q/\Delta T = 30\ 000)}$	0.39	.42	0.915
using log scale (Figure 3.24)			
$C_{(Q/\Delta T = 9217)} = 0.696$	$C_{(Q/\Delta T = 8572)} = .44$		1.1
Approx. area = 64.9 m²	Approx. area	78.33	59.57 m²

BUDGET PRICE $= Q/\Delta T \times C$

$= 9217 \times 0.696$

$= £6416$

BUDGET PRICE   £3722   £9306

A finned tube double pipe is the choice here.

3.6.2. Process water cooler

Example—A heat exchanger is required to cool process water flowing at 85 kg/s from a temperature of 120°C to a temperature of 40°C using cooling water at the same flow rate with an inlet temperature of 20°C and an outlet temperature of 100°C. The pressures of the two fluids are around 5 bar.

In this case the application is suitable for plate exchangers, shell-and-tube exchangers and double pipe exchangers.

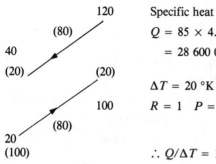

Specific heat water = 4.2

$Q = 85 \times 4.2 \times (120 - 40) = 28\,600$ kW

$= 28\,600\,000$ Watt

$\Delta T = 20\,°K$

$R = 1 \quad P = 0.8$ Impossible multi-pass operation (see Figure 3.2). Single pass mandatory.

$\therefore Q/\Delta T = 1.43 \times 10^6$

Since $Q/DT > 1 \times 10^6$ use C factor $Q/\Delta T = 1 \times 10^6$

	Shell and Tube	Plate	Double Pipe Finned	Plain
U	= 938	4171	325	938
C	= 0.047	0.015	0.12	0.139
area	= 1524 m²	342 m²	4400 m²	1524 m²
BUDGET PRICE	£67 200	£21 450	£171 600	£198 770
	Prices quoted for equipment by a manufacturer.	£19 000/32 000 for ΔP of 10 MWG/1 MWG 1/0.1 Bar.		

Clearly all conditions allowing for a gasketed plate exchanger the obvious choice must be this.

3.6.3. Light hydrocarbon condenser

Example—A light hydrocarbon vapour is to be condensed under saturation conditions at a temperature of 120°C. It enters the condenser as saturated vapour and leaves as saturated condensate. It is flowing at a rate of 300 kg/s and has a latent heat of condensation of 200 kJ/kg. The coolant is treated cooling water entering the system at 20°C and leaving and 50°C.

In this case for a condensing unit, a plate heat exchanger would be unsuitable and the choice lies between the double pipe heat exchanger and the shell-and-tube exchanger.

120°C

(100) —————◄——— (70)
          50

20 ————————►

$$Q = 300 \times 200$$

$$= 60\ 000\ \text{kJ/s}$$

$$= 60\ 000\ 000\ \text{Watt}$$

$$\Delta T_{\text{LM}} = \frac{30}{\log_e 1.428} = \mathbf{84.1\ °K}$$

$$F_{\text{t}} = 1.0 \quad \therefore\ \Delta T = 84.1\ °K$$

$$Q/\Delta T = 713\ 350$$

| | | **Double Pipe** |
| **Shell and Tube** | **Finned** | **Plain** |

$C_{(Q/\Delta T = 10^5)}$	= 0.094	$C_{(Q/\Delta T = 10^5)}$	= 0.15		= 0.17
$C_{(Q/\Delta T = 10^6)}$	= 0.058	$C_{(Q/\Delta T = 10^6)}$	= 0.13		= 0.17
C	= 0.06	C	= 0.132	C	= 0.170
U	= 764	U	= 307	U	= 764
area	= 933 m²	area	= 2323 m²	area	= 933 m²

BUDGET PRICE	£42 800	£94 162	£121 269

The choice is therefore a shell and tube exchanger.

Let us assume that the process is corrosive and the material selected was type 316 stainless steel, the design pressure is 40 bar and a removable bundle with straight tubes is required.

(1) From Figure 3.4 type of exchanger is BES.

(2) From cost curves, Figure 3.6a–h—tubes assumed at 6096 mm, area 933 m²

From 3.6b price/m² = 44 from 3.6a dia > 46″ (therefore 2 shells in series).

Factor for type Figure 3.6d = 1.165

Factor for pressure Figure 3.6f = 1.43

Factor for material (316) Figure 3.6h = 2.11

BUDGET PRICE = 935 × 44 × 1.165 × 1.165 × 1.43 × 2.11

= £144 300   for a stainless steel BES floating head exchanger, design pressure 40 bar.

3.6.4. Comparison of air cooled heat exchanger and cooling tower

Example—A hydrocarbon is to be condensed at 150°C with a condensing heat load of 10 MW. Compare the approximate cost of an air cooled heat exchanger (air inlet temperature 20°C, outlet temperature 40°C) which condenses the vapour directly with those of a shell-and-tube heat exchanger which condenses the vapour, combined with a package cooling tower for heat rejection. Assuming a cooling water inlet temperature of 25°C and outlet temperature 45°C.

Air cooler

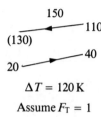

$$\Delta T = 120 \text{ K}$$

Assume $F_T = 1$

$$\frac{Q}{\Delta T} = 83.3 \text{ kW/K}$$

From table at 30 kW/K $C = 0.6$
100 kW/K $C = 0.35$

$\therefore$ C by calc. $= 0.388$

Budget Price $= £32\ 300$

Combined shell and tube cooling tower

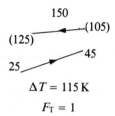

$$\Delta T = 115 \text{ K}$$

$$F_T = 1$$

$$\frac{Q}{\Delta T} = 87\ 000$$

From table at 30 kW/K $C = 0.16$
100 kW/K $C = 0.094$

$\therefore$ C by calc. $= 0.1016$

Exchanger cost: £8 841

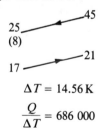

$$\Delta T = 14.56 \text{ K}$$

$$\frac{Q}{\Delta T} = 686\ 000$$

From table at 30 kW/K $C = 0.048$
100 kW/K $C = 0.029$

$\therefore$ C by calc. $= 0.0311$

Tower cost: £21 356

Total cost $= £30\ 197$

In this case, the shell-and-tube heat exchanger/cooling tower combination is cheaper, but not significantly so compared to the accuracy of this type of calculation. The cost of fan operation should be borne in mind and also the cost of the cooling water loss due to evaporation and the cost of cooling water pumping.

4. Application, Problems and Examples

4.1. Introduction

4.2. Crude Pre-heat Train

4.2.1. Process description

4.2.2. Data extraction and energy targeting

4.2.3. Pinch design

4.2.4. Design evolution

4.2.5. Design evaluation

4.2.6. Conclusions

4.3. Aromatics Plant

4.3.1. Introduction

4.3.2. Process description

4.3.3. Data extraction and energy targeting

4.3.4. Pinch design

4.3.5. Process design considerations

4.4. Evaporator/Dryer Plant

4.4.1. Process description

4.4.2. Data extraction and energy targeting

4.4.3. Heat pumping strategy

4.4.4. Evaluation of heat pumping schemes

4.4.5. Conclusions

4.1. Introduction

The purpose of this section of the Guide is to illustrate the application of the integration techniques in "real-life" case studies. All of the case studies described are based on engineering designs performed in industry. Of necessity, many details are omitted. However, it is hoped that the material will help the reader to appreciate the use of integration techniques in context. There are three studies selected to illustrate the breadth of applicability of the techniques, from the small and simple to the large and complex. Hopefully, out of the studies presented, the reader will recognise one or more as familiar territory.

4.2. Crude Pre-heat Train

The fractionation of crude oil into its various crude "cuts" is a common process, standing "at the beginning" of the utilisation of oil. In the fractionation process to be described here, a facility needed uprating to handle increased demand. Design studies were carried out by a contractor for the operating company, the conclusions of which were that it was not possible to increase the throughput of the plant without installing a new fired heater. This was seen as the cheapest capital option and was therefore suggested. However, the operating company decided that this was an undesirable course, not only because of energy considerations, but also because of operability and safety problems inherent in the proposed system. The (then newly-developed) integration techniques described in this Guide were applied in order to identify quickly a more attractive option.

4.2.1. Process description

The contractor's proposed flowsheet is shown in Figure 4.1(a), in simplified form. The crude oil feed stream is pre-heated in three sections by interchange with the hot fractions returning from the distillation columns. The first section runs from storage to a desalter unit, the second from the desalter to a preflash column which separates out some light naphtha, and the third from the bottom of the preflash to the crude tower. Process heating is provided by a fired heater, which pre-heats the crude into the crude tower and provides reboiling for the stripper. The proposed additional fired heating was to have taken the form of a hot oil circuit placed immediately upstream of the existing heater. Layout constraints meant that the heater for the proposed oil circuit would have to have been placed away from the main plant. In fact it would have to have been sited on the opposite side of a busy site road. Imagine the operability and safety problems! However, the contractor's design achieved more or less full utilisation of the existing heat interchange equipment, although the large air-cooled heat exchanger on the column overheads would have needed modification.

4.2.2. Data extraction and energy targeting

The heat exchanger network shown in flowsheet form in Figure 4.1(a) is reproduced in grid form in Figure 4.1(b), with stream temperatures and match heat loads marked on. Because of the project time pressures, it was not possible to do a detailed computer simulation of the stream T/H profiles. Instead, data was extracted from the contractor's flowsheet in the manner described in Section 2.2 of the Guide, using the given design heat loads and temperatures. This gave the stream data listed in Table 4.1.

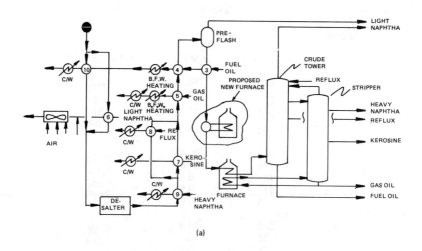

(a)

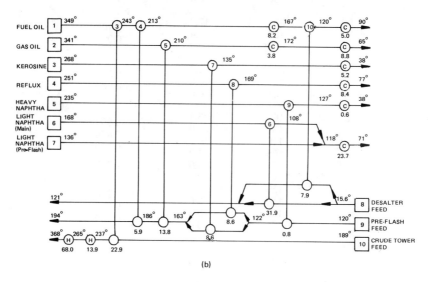

(b)

Figure 4.1—Base case design

The smallest values of ΔT_{min} in the contractor's design are 7°C at the cold end of match 9, and 13°C at the cold end of match 7. However, ΔT_{min} values in other matches are much higher than this. Hence it was decided, for the "first look" at energy targeting to take a global ΔT_{min} value of 20°C. Calculating the Problem Table on this basis gave a hot utility requirement of 60.7 MW. The plant *prior to uprating* was consuming 68.0 MW, and the contractor's proposals required an extra 13.9 MW, *i.e.* a total heat input of 81.9 MW! Hence the calculated target indicated a potential saving of about 35%.

193

Table 4.1

Stream	T(°C)	H(MW)	$CP = (\Delta H/\Delta T)$
		Data	
1	349	49.8	0.215
	243	27.0	0.197
	213	21.1	0.178
	167	12.9	0.168
	90	0.0	
2	341	26.4	0.105
	210	12.6	0.100
	172	8.8	0.087
	111	3.5	0.076
	65	0.0	
3	268	13.9	0.065
	135	5.2	0.054
	38	0.0	
4	251	17.0	0.105
	169	8.4	0.091
	77	0.0	
5	235	1.4	0.008
	127	0.6	0.007
	38	0.0	
6	168	43.1	0.600
	136	23.9	0.478
	118	15.3	0.410
	108	11.2	0.303
	71	0.0	
7	136	12.6	0.256
	118	8.0	0.210
	108	5.9	0.159
	71	0.0	
8	15.6	0.0	0.379
	121	39.9	
9	120	0.0	0.400
	122	0.8	0.422
	163	18.1	0.600
	186	31.9	0.725
	194	37.7	
10	189	0.0	0.477
	237	22.9	0.406
	265	36.8	0.660
	368	104.8	

From this point in the study, the prospects for finding a revamp which avoided using the extra fired heater appeared very good indeed and provided a tremendous stimulus to the operating company's and the contractor's engineers to find such a design.

4.2.3. Pinch design

Calculating the Problem Table (as in Section 2.2.2 of the Guide) for $\Delta T_{min} = 20°C$, a heating requirement of 60.7 MW, a cooling requirement of 42.5 MW, and a pinch at 173°C are obtained. On inspection, it can be seen that this pinch is caused by the onset of vaporisation in the crude feed stream, 9. We now commence a design using the "Pinch Design Method", keeping in mind that we want to maximise compatibility with the existing plant.

(i) Above the pinch—Figure 4.2(a) shows the stream set above the pinch. The first point to realise is that, because there are five hot streams and only one cold stream at the pinch, the cold stream must be split five ways! This would clearly be impractical, so one or more of these splits must be evolved out. As a first simple evolution the load on stream 5 (the heavy naphtha), above the pinch which is small in comparison with the net heating duty (less than 1%) can be ignored in the above-the-pinch design, reducing the required number of stream split branches on stream 9 to four.

The first design decisions are shown in Figure 4.2(b), with stream 5 removed and the load on the process heater adjusted accordingly. Note that the four matches against the split stream 9 (the feed stream to the preflash tower) are basically present in the existing plant (although not against so many split branches). Having decided on these matches, match number 3 is added because it already exists. We now assign loads to

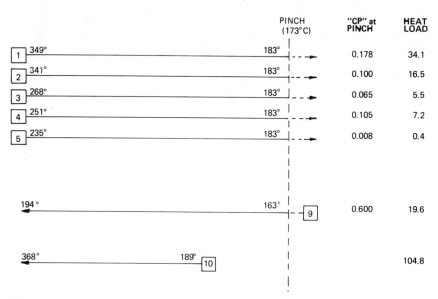

Figure 4.2(a)—Stream set above the Pinch

195

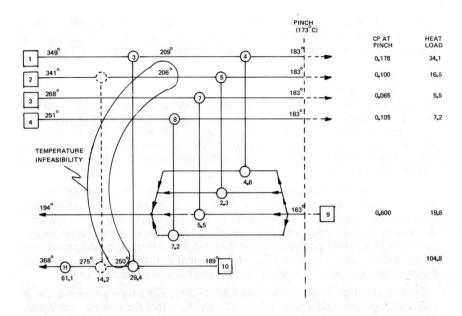

Figure 4.2(b)

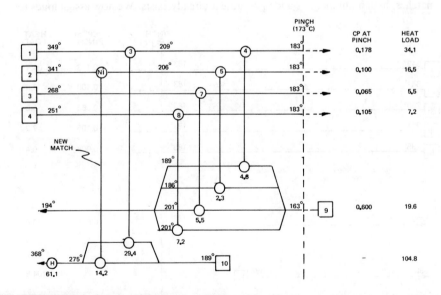

Figure 4.2(c)

Figure 4.2(b and c)—Above-the-Pinch design

these matches. The loads on matches 7 and 8 are maximised to "tick-off" streams 3 and 4. Match 3 should not cool stream 1 below 209°C (because stream 10 supply temperature is 189°C). This dictates the maximum load on match 3, which becomes a design decision. If stream 1 is cooled to 209°C in this match, the load on match 4 is fixed. This in turn fixes the load on match 5 by enthalpy balance on stream 9. Fixing the load on match 5 allows us to calculate the temperature of stream 2 on the hot side of the match. It comes out to be 206°C which means that a residual cooling duty of 14.18 MW is left on stream 2. Because we have obeyed the pinch design method, this is exactly the size of "hole" that we find is left to be filled on stream 10 (having placed the minimum load of 61.1 MW on the fired heater). Hence a match of this load is required between stream 2 and 10, as indicated by the dotted line in Figure 4.2(b). However, a sequential match as indicated by Figure 4.2(b) is not feasible. A possible way to solve this problem is to split stream 10 and to place match 3 and the new match on parallel branches, as shown in Figure 4.2(c).

This is our completed above-the-pinch design. Note that it only requires one basic new match compared with the existing plant.

Finally, note that there is a small violation of ΔT_{min} at the cold end of the new match in Figure 4.2(c). However, remember that the ΔT_{min} of 20°C was chosen arbitrarily in the first place.

(ii) *Below the pinch*—Figure 4.3(a) shows the stream set below the pinch. Once again, the four way split of stream 9 is required (by the "*CP*-rules"—see Section 2.2 of the Guide), yielding the same topological set of four matches, 4, 5, 7, 8. The other existing matches 6, 9, 10 and all the coolers are also included as shown in Figure 4.3(b). If we assign the "base case" loads to matches 6, 9 and 10 (*i.e.* the matches away from the pinch), then the loads on the pinch matches 4, 5, 7 and 8 turn out as shown. Notice that because the sum of the *CP*s of streams 1 to 4 is almost exactly equal to the *CP* of stream 9 at the pinch, minimum driving force is maintained throughout the pinch matches. This means that there is no flexibility in choice of branch flow rate in the split stream 9.

(iii) *Complete MER design*—The completed MER design is obtained by merging the two systems "above" and "below", with the result shown in Figure 4.4(a). The split stream 9 branch flows calculated in the two halves are not compatible. However, because some flexibility exists in the above-the-pinch splits, the branch flows calculated for below the pinch are adopted for the combined design. This then means that the target temperatures on the individual branches are changed in the combined design (compare Figure 4.2(c)). However, they remain feasible against the hot stream temperatures.

4.2.4. Design evolution
The MER design shown in Figure 4.4(a) achieves a 10% energy saving over the *existing* plant and a 25% saving over the contractor design. Topologically, the only difference between the MER design in Figure 4.4(a) and the existing plant is one new match (labelled "N1"). The design therefore appears a promising starting point for the evolution of a good revamp scheme which avoids the need for supplementary heating.

The next step is to carry out a "*UA*-analysis" as described in Section 2.2 of the Guide. Values for UA (= $Q/\Delta T$) for the contractor's "base case" design and for the

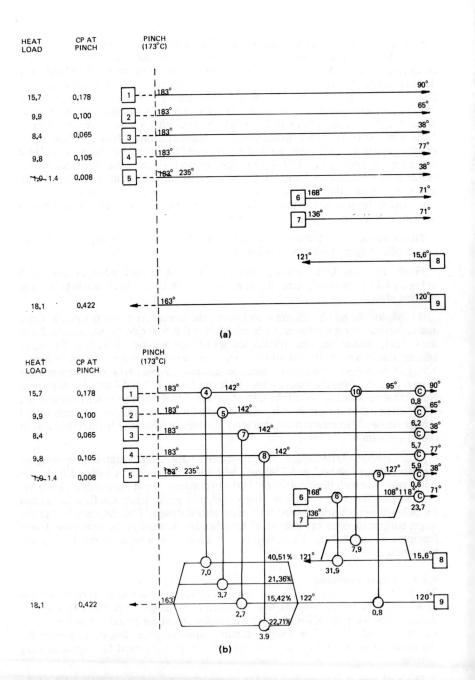

Figure 4.3—Below-the-Pinch design

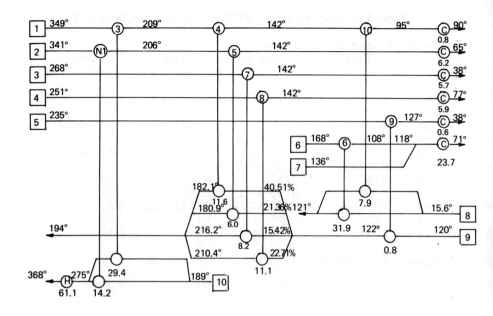

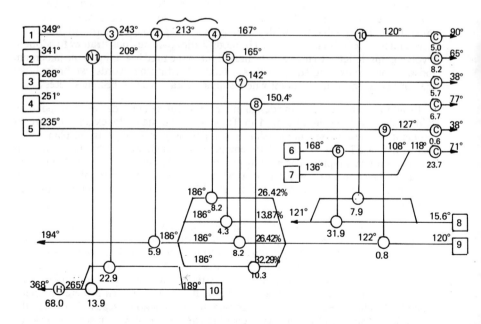

Figure 4.4—MER design and first evolution

Table 4.2—Comparison of "UA" (= $Q/\Delta T_{LM}$) Values (in MW/°C)

Match	Contractor Design	MER Design	Evolved Design (4-way split)	Evolved Design (3-way split)
N1	—	0.393 (NEW)	0.332 (NEW)	0.332 (NEW BUT RE-USE 5)
N2	—	—	—	0.210 (NEW)
3	0.288	0.714 (MOD)	0.337	0.337
4	0.159	0.549 (MOD)	0.412 (MOD)	0.476 (MOD)
5	0.152	0.286 (MOD)	0.147	—
6	0.462	0.462	0.462	0.506
7	0.196	0.293 (MOD)	0.198	0.193
8	0.132	0.454 (MOD)	0.241 (MOD)	0.234 (MOD)
9	0.022	0.022	0.022	0.022
10	0.111	0.180 (MOD)	0.111	0.111
Air Cooler	0.550 (MOD)	0.550 (MOD)	0.550 (MOD)	0.392
ΣUA	2.072	3.903	2.812	2.813

synthesised MER design are shown in columns 2 and 3 of Table 4.2. Values are given for the interchangers and for the big cooler (the air-cooled heat exchanger on stream 6/7). It can be seen that the MER design pays a heavy penalty in terms of additional area and number of matches in need of modification. An obvious strategy to adopt in evolving this design is to increase the heat input to the fired heater up to the maximum possible on the existing equipment, *i.e.* 68.0 MW. In other words, we "relax" the design just as far as to the point where supplementary heating becomes necessary.

By adopting this strategy, the design in Figure 4.4(b) is obtained. The corresponding *UA* values are listed in column 4 in Table 4.2. The match loads and the stream temperatures are chosen for maximum compatibility with the existing plant. This allows matches 3, 5, 6, 7, 9 and 10 to remain un-modified. Match 4 has been split into two parts for easy piping. The existing part of match 4 is left on the hot end of the full stream 9. The new part is situated on a split branch. Notice that not only is the extent of plant modification reduced in the design of Figure 4.4(b), but also the amount of additional area.

Two potential problems with this design remain, however. These are the 4-way split of stream 9 and the need for an expensive modification of the big air cooler. This last difficulty can be overcome by adding area to matches 6 and 4, allowing load to be shifted round the loops from the air cooler to any of the water coolers on streams 2, 3 and 4. The effect of eliminating one of the stream split branches is shown in Figure 4.5(a). The branch chosen is that carrying match 5, with match 5 therefore being completely eliminated. Match 5 is chosen for elimination because it carries the smallest load amongst the branch matches (and hence its removal causes least upset of driving forces amongst the remaining matches) and because the existing shell of match

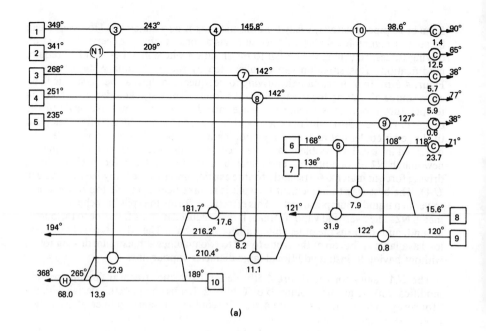

(a)

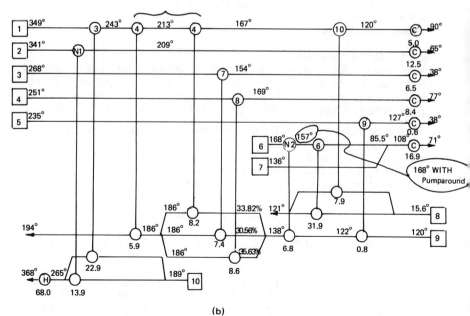

(b)

Figure 4.5—Second design evolution

5 can be re-piped relatively easily for utilisation in the new match N1. However, it can be seen in Figure 4.5(a) that this decision has a considerable "squeezing" effect on driving forces, returning the design to the situation where matches 7 and 10 require modification. This situation can be alleviated by the evolutionary step shown in Figure 4.5(b). In order to ease the driving force squeeze in the matches to stream 9, an extra source of heating for stream 9 must be found. The only candidate available (on temperature grounds) is stream 6. Hence introducing new match N2 leads to the design in Figure 4.5(b), with matches 7 and 10 restored to their un-modified state, and the large air cooler not now needing modification. Match 6 however does have to be increased in size. This problem can be overcome by splitting stream 6 and placing the new match N2 on one branch, and match 6 on the other. The effect is to maximise driving force in match 6 sufficiently for the existing unit to cope (only 10% increase in UA). The best way to implement the split is to take heat from the top of the crude tower in a liquid "pump-around". This effectively forms the split branch for the new match N2. The overhead vapour stream is then reduced in mass flow due to the pump-around, and so forms the other split branch for match 6. The effect is to split the heat load available at the top of the crude tower to two matches at maximum driving force, without having to install additional large diameter vapour lines.

The UA values for the Figure 4.5(b) design (with the stream 6 "pump-around" modification) are given in column 5 of Table 4.2. Let us compare this design with the "four-way split" design of Figure 4.4(b). It requires the same number of exchanger modifications and the same amount of area. It requires one more match but one less stream split. At this stage of the study it is not possible to say which design is better. However, what can be said is that two promising candidate designs have been found as alternatives to the contractor's design.

4.2.5. Design evaluation

Next in any study, a more detailed check of equipment performance and system operability is required, followed by engineering design specification and costing. However, a preliminary "feel" for heat exchanger costs can be gained by using the tables in Section 3 of the Guide, as illustrated for this case study in Table 4.3.

After detailed evaluation, the design chosen for construction was that shown in Figure 4.5(b). It is shown in flowsheet form (including the new pump-around) in Figure 4.6. The techniques described in this Guide saved energy to a value of about £1 million per year. In addition the design was safer (due to the elimination of a fired heater) and, perhaps surprisingly, more operable. The presence of the three-way split

Table 4.3—Marginal $U.A$ Values and Costs

Match	$\Delta(U.A)$ $(\equiv Q/\Delta T_{LM})$ (W/K)	C (£/(W/K))	C_T (£)
N1	1.80×10^5	0.104	18 700
N2	2.10×10^5	0.102	21 400
4	3.17×10^5	0.098	31 100
8	1.02×10^5	0.110	11 200
			82 400

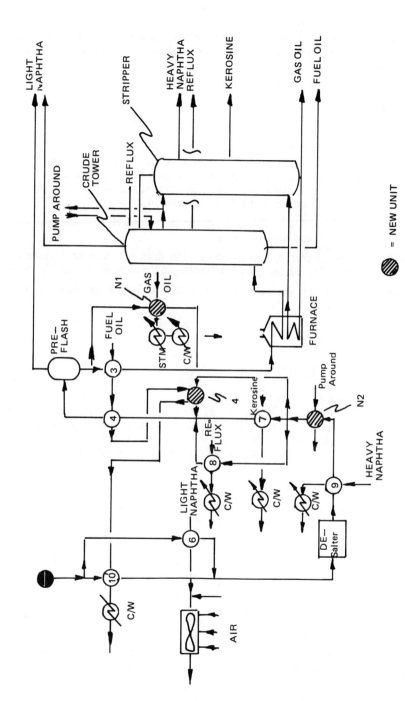

Figure 4.6—Final uprating scheme

203

on stream 9 meant that if the crude feedstock was changed, yielding a different balance between light and heavy fractions, the branch flows could be adjusted to compensate.

The design in Figure 4.5(b) is now installed and is operating satisfactorily, achieving the expected rates.

4.2.6. Conclusions
The main points highlighted by this study are:

(i) The Pinch Design Method can be used systematically to produce good "revamp" designs, even where the existing heat exchanger network is complex. It allows a productive interaction with the engineer's experience (a good example is the use of the pump-around in the preferred solution).

(ii) Designs produced by proper use of the method are elegant, sometimes yielding both energy and capital savings.

(iii) A higher degree of process integration does not necessarily cause control problems. If the integration is well balanced the controllability can be enhanced.

(iv) Parallel stream splitting is a practical tool for improving energy recovery and operability.

(v) The Pinch Design Method can be employed to give good designs in rapid time and with minimum data.

4.3. Aromatics Plant

4.3.1. Introduction
The plant concerned in this study is part of one of the largest aromatics complexes in Europe. It was commissioned in 1969 and uses conventional technology. It can be considered typical of state-of-the-art aromatics plant design. The feedstock is a naphtha containing chiefly paraffins and cycloparaffins which are reformed into product containing paraffins and aromatic compounds.

4.3.2. Process description (Figure 4.7)
A heart cut of naphtha from which both light and heavy ends have been stripped is the principal plant feedstock. The feed is vaporised (H1) and passed through a desulphurisation reactor (R1). Heat is recovered from the reactor effluent in two interchangers (A, B) prior to condensation (C1) and gas separation (F1). The liquid from the separation stage is re-heated by reactor effluent (B) and fed to a stripping column (D1) in which the light ends and sulphur-containing compounds are removed.

The desulphurised naphtha stream from the column is mixed with recycle gas. The two phase mixture is preheated in a series of process interchangers (D, C). The mixture is finally raised to the reaction temperature of 500°C by a radiant furnace (H2) fired by a mixture of gas and fuel oil.

The reactions take place in a series of reformers (R2, R3) with intermediate fired heating (H3). The reformer effluent, at 490°C, is cooled in a series of interchangers,

204

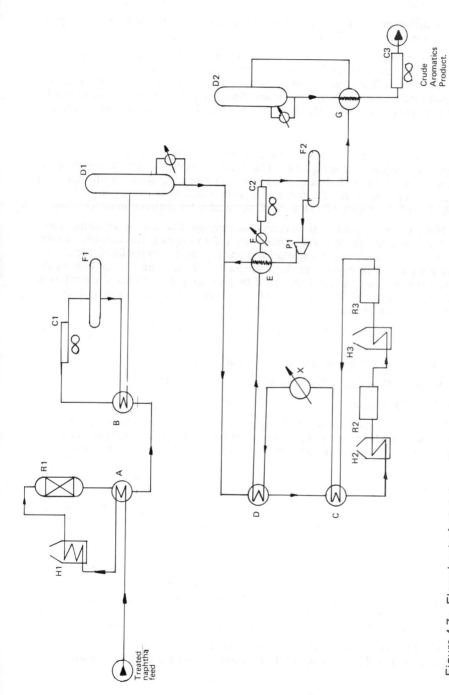

Figure 4.7—Flowsheet of naphtha reformer prior to modification

some preheating the feed (C, D, E) and some providing the heat source for other process requirements (F, X).

Final cooling and gas separation takes place in C2 and F2. The gas recycle is compressed (P1) and preheated (E) prior to mixing with the liquid reformer feed. The liquid from the flash drum is passed to a column for stabilisation (D2) and a conventional feed/tails interchanger (G) is installed to reduce the reboil requirement by adding feed preheat. The reformate stream is finally cooled in C3 prior to storage.

4.3.3. Data extraction and energy targeting
The "base case" heat exchanger network (shown in flowsheet form in Figure 4.7) is represented in grid form in Figure 4.8. The grid shows all the flowsheet heat exchangers, heaters and coolers with their heat loads and corresponding stream temperatures (in °C). The smallest observed approach temperature in the heat exchangers is 10°C, at the hot end of exchanger B. Thus, 10°C was used as the initial estimate for ΔT_{min}.

Many of the streams in the process have pronounced non-linear temperature/enthalpy profiles (due to partial condensation, for example). This non-linearity was represented by temperature/enthalpy data points generated by modelling the streams using a physical properties computer program. For the present description of the study computation has been simplified by generating data points as described in Section 2.2 of the Guide, using the known flowsheet heat loads and temperatures only. This linearised data is shown in Table 4.4.

Using the base case data and $\Delta T_{min} = 10°C$ the Problem Table gives the following results

Minimum Hot Utility Requirement   = 46.5 tonne cal/hr

Minimum Cold Utility Requirement   =   8.9 tonne cal/hr

Pinch Temperature   = 145°C

The current hot utility usage is 57.2, representing a 23% excess energy usage above the minimum. In a high tonnage process with high energy usage this meant a considerable financial incentive to identify schemes to achieve part, if not all, of this saving!

4.3.4. Pinch design
Having used the Problem Table Method (Section 2.2.2.3) to establish the pinch temperature at 145°C, it is a simple matter to identify the stream sets for "above the pinch" and "below the pinch" (Figures 4.9(a) and 4.9(b)). Notice that in this process the pinch is caused by stream 4 starting at 140°C with a relatively high heat capacity flowrate i.e. 0.2 tonne cal/hr °C.

Having established the above and below the pinch stream sets one can now identify a design to achieve maximum energy recovery. Remember, however, the "golden rule" of retrofit heat exchanger network design—always aim to re-use as much of the existing hardware as possible in your minimum energy design, even if this means introducing otherwise unnecessary stream splits and/or ΔT_{min} violations!

(i) Above the pinch—From Figure 4.9(a) we see that streams 2, 6 and 9 each require to be cooled to 150°C with one of the four cold streams at the pinch (1, 3, 4 or 8). For

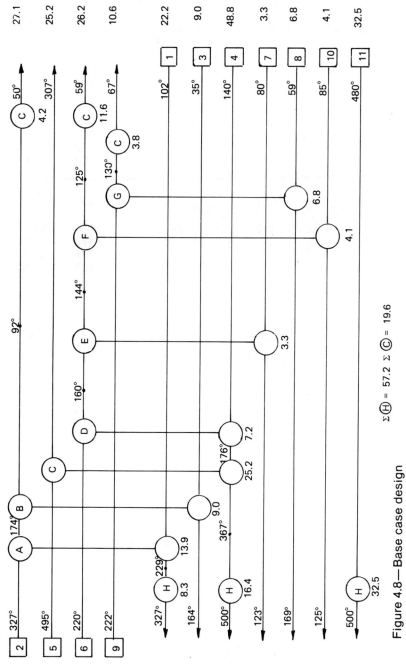

Figure 4.8—Base case design

207

Table 4.4—Process Stream Data

STREAM NO	SUPPLY TEMP. (°C)	TARGET TEMP. (°C)	TEMP./ENTHAPLY DATA T (°C)	H (tonne cal/hr)
1	102	327	102	0.0
			229	13.9
			327	22.2
2	327	50	327	27.1
			174	13.2
			92	4.2
			50	0.0
3	35	164	35	0.0
			164	9.0
4	140	500	140	0.0
			176	7.2
			367	32.4
			500	48.8
5	495	307	495	25.2
			307	0.0
6	220	59	220	26.2
			160	19.0
			144	15.7
			125	11.6
			59	0.0
7	80	123	80	0.0
			123	3.3
8	59	169	59	0.0
			169	6.8
9	220	67	220	10.6
			130	3.8
			67	0.0
10	85	125	85	0.0
			125	4.1
11	480	500	480	0.0
			500	32.5

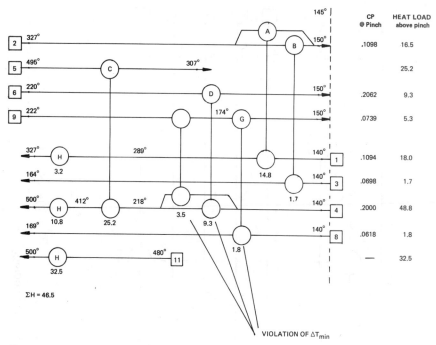

	CP @ Pinch	HEAT LOAD above pinch
	.1098	16.5
		25.2
	.2062	9.3
	.0739	5.3
	.1094	18.0
	.0698	1.7
	.2000	48.8
	.0618	1.8
	—	32.5

VIOLATION OF ΔT_{min}

Figure 4.9(a)—Above-the-Pinch design

temperature feasibility each one of these matches must have $CP_{HOT} \leqslant CP_{COLD}$. Thus, it seems clear that stream 2 should be matched with stream 4, stream 6 should be split and matched with streams 4 and 8, and stream 9 should be matched with stream 1.

For a new design these would almost certainly be the essential pinch matches. For a retrofit study we must think again, since this design would render matches A, B and G redundant.

The heat in stream 2 is recovered through matches A and B, with the requirement for a new stream split. The heat in stream 6 is recovered through match D (even though the match will now inevitably violate ΔT_{min}, due to the CP_{HOT} being greater than CP_{COLD}) and the heat in stream 9 is partially recovered by stream 8 (again with an inevitable ΔT_{min} violation) *via* match G. Following these design decisions, and "ticking off" of streams, the design problem is now reduced to that of recovering the heat in stream 5 and the residual heat in stream 9 (between 222°C and 174°C).

The available heat sinks are streams 1, 4 and 11. Examination of these sinks quickly shows that the only candidate sink for the heat in stream 9 is stream 4, and only then after a stream split to bypass exchanger D! This new match violates ΔT_{min}. Because of its small driving forces and the requirement for a stream split it is hardly surprising that the existing design misses this match!

The remaining design problem, *i.e.* heat recovery from stream 5, is easily solved

209

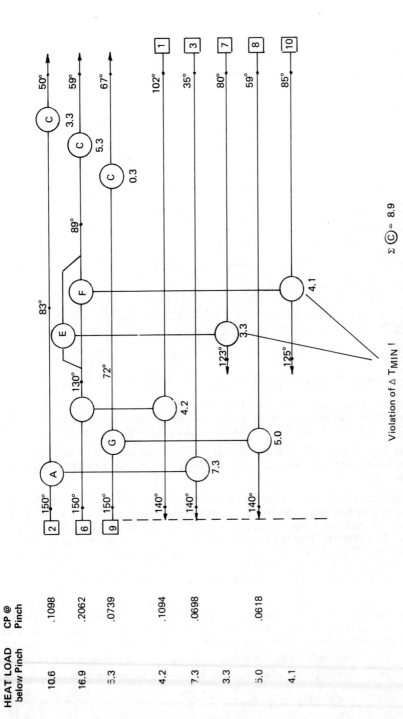

Figure 4.9(b)—Below-the-Pinch design

$\Sigma \, \text{C} = 8.9$

Violation of ΔT_{MIN}!

HEAT LOAD below Pinch	CP @ Pinch
10.6	.1098
16.9	.2062
5.3	.0739
4.2	.1094
7.3	.0698
3.3	
5.0	.0618
4.1	

210

using the existing match C. The residual heating requirements (total load = 46.5) are supplied by utility heaters.

(ii) *Below the pinch*—Streams 1, 3 and 8 in Figure 4.9(b) must be heated to 140°C by process interchange. Existing matches A and G (also used above the pinch) are obvious, feasible design choices, leaving the requirement for a second new match between streams 6 and 1. These three matches satisfy the heating requirements of streams 1, 3 and 8. Streams 7 and 10 are now supplied with heat *via* existing matches E and F with stream 6. Again, stream splitting is essential for feasibility, but even so matches E and F now violate ΔT_{min}.

There are alternative designs to that shown in Figure 4.9(b) which would not violate ΔT_{min}. However, these designs would be less compatible with the base case.

(iii) *Complete design and energy relaxation*—At this stage in the study, having generated separate above and below the pinch designs, detailed evaluations of the feasibility of stream splitting and of increasing the surface area of existing units was undertaken. It was decided that stream splitting would not be attractive and the design was relaxed by importing more energy. Apart from eliminating stream splits, this relaxation aimed to increase ΔT_{min} (and decrease surface area requirements in general) and to merge the duplicate matches A and G.

The resulting design, labelled Phase III, is shown in Figure 4.10. The hot utility usage of 52.1 corresponds to 12% above minimum. Further energy relaxation to eliminate the new match between streams 6 and 1, and also to further minimise the additional surface area required in existing matches, leads to the design labelled Phase II in Figure 4.11. This phase of the project (along with an earlier non-heat integration energy saving idea—Phase I) is now being installed.

Phase III represents an "add-on" energy saving project which is fully consistent with Phase II. It is currently being designed and costed in more detail. Thus, by using the pinch design method, followed by energy relaxation, we have established a strategy for the phased improvement of the aromatics process heat recovery! This is a feature of the techniques described in this Guide which is very important, particularly in an economic climate where capital for projects is restricted. Phases II and III and the minimum energy design are compared with the base case in Table 4.5.

4.3.5. Process design considerations

During the energy relaxation phase of the project it became obvious that a simple process change could be implemented which would further reduce the energy requirement to around 7% above the minimum. At this time, this process change is

Table 4.5—Aromatics study—Summary

Scheme	H	% ABOVE MIN
Base Case	57.2	23%
Phase II	54.7	18%
Phase III	52.1	12%
Minimum Energy	46.5	0%

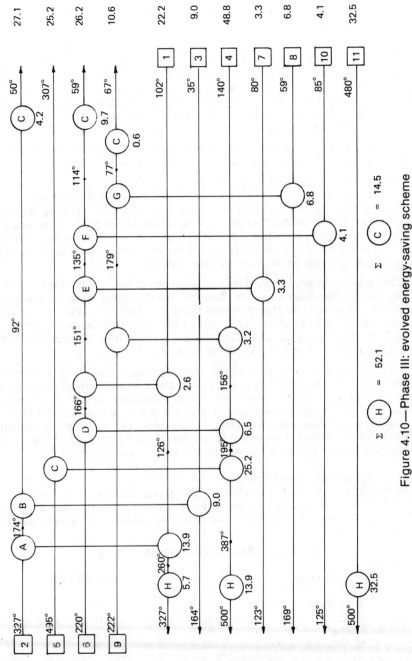

HEAT LOAD

Figure 4.10—Phase III: evolved energy-saving scheme

$$\Sigma \, \bigcirc\!\!\!\!\text{H} \; = 52.1$$

$$\Sigma \, \bigcirc\!\!\!\!\text{C} \; = 14.5$$

212

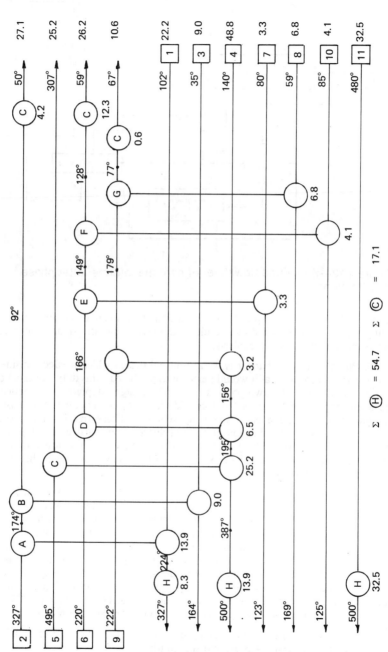

Figure 4.11—Phase II: evolved energy-saving scheme

213

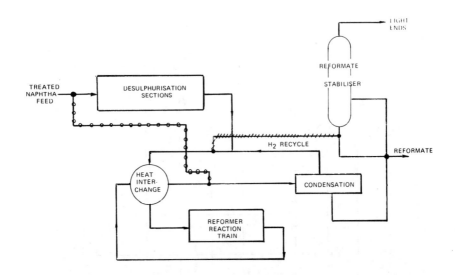

Figure 4.12—Simplified flowsheet of naphtha reformer after phased
modifications

the subject of a patent application and further details cannot be given here. This
change came to light as a direct result of the systematic and rigorous procedures
necessitated by the network analysis.

The Phase II heat exchanger is of an unusual design which needed considerable
design effort to ensure that it would perform as predicted. This duty consisted of the
partial vaporisation of a two phase stream. The design adopted was a vertical rod
baffle exchanger with vapour belt and shellside liquid injection nozzles. The constraints
of plant layout forced the use of a vertical exchanger. Thereafter, it was necessary to
consider pressure drop, vibration and liquid distribution, particularly slugging and
their effect on the process design. The complexity of the resulting design reflects the
lengths to which the designers had to go to in order to have confidence that the duty
would be satisfactorily achieved.

In many circumstances, the design team when faced with such a problem, might
choose to overcome the problem by altering the heat exchanger match without a
quantitative knowledge of how this would affect the energy recovery network. The
techniques of heat exchanger network design not only indicate the appropriate matches
to achieve minimum energy but also quantify the energy penalty involved in pursuing
other network configurations. Thus there is a definite incentive for the process
designer to achieve the optimum network design. This incentive only existed as a
qualitative feeling prior to the introduction and development of the techniques
introduced in this Guide. Figure 4.12 shows the plant after the phased modifications
have been introduced. It is compatible with the grid representation of Figure 4.10.

The annual savings generated by this project were in the order of £0.5 million *p.a.*
and the payback achieved was around six months.

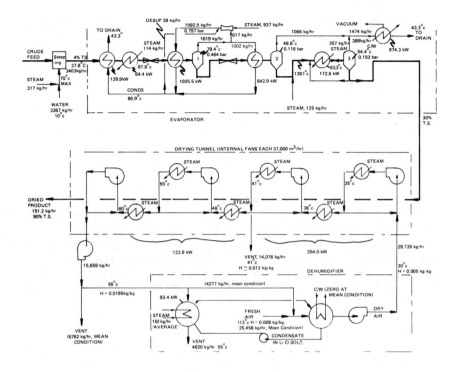

Figure 4.13—Base case design

4.4. Evaporator/dryer plant

This study relates to a survey carried out in a food processing factory. After first stage processing, the product is fed in dilute aqueous solution to an evaporation and drying section. The latter section contains modern, continuous operation equipment, but the first stage processing comprises old and inefficient batch equipment. It had been decided to rebuild completely the "front end" of the process, and so the opportunity was taken to perform an energy utilisation study to see whether efficiency improvements could be made in the total system, with any hardware changes to be made whilst the plant was shut down.

4.4.1. Process description

The continuous operation evaporator/dryer section is shown in Figure 4.13. Crude feedstock is extracted by adding hot water. The extract solution (4% total solids) is fed into the continuously-operated evaporator/dryer plant. The evaporator has three stages, the first two operating on the "multiple effect" principle. Flash vapour from the first stage drives the second stage. Also, some vapour from the first stage is recompressed by a steam ejector for re-use in driving the first stage (heat pumping principle). The steam ejector and the third stage are driven by utility steam (9.29 bar,

215

saturated). The flash vapour from the second and third stages is condensed against cooling water in a vacuum condenser. The evaporator produces liquor of 30% total solds which is fed to the dryer. The dryer operates with a continuous belt, carrying the product through seven temperature zones ranging from 30°C to 60°C. Internal drying tunnel air circulation is maintained by four fans, and heat input is via six steam radiators (using 6.53 bar utility steam). Wet air from the dryer is vented at the mid-point (41°C) and from the end of the drying tunnel (60°C), but depending on ambient conditions, a controlled amount of the hottest wet air is recycled. This, together with fresh air make-up is dehumidified by contact with lithium chloride solution down to a humidity of 0.005 kg/kg at a controlled temperature of 30°C. The lithium chloride solution is regenerated by spraying over steam heating coils in a "return air" (55°C) stream.

4.4.2. Data extraction and energy targeting
The hot and cold stream data extracted from the flowsheet in Figure 4.13 are given in Table 4.6, divided into the two broad areas of the evaporator and the dryer.

In the evaporator area, only those vapour loads which arise directly because of process duties are included. This means that at this (energy targeting) stage, we ignore the steam ejector (which is a heat pump), and stick to analysis of the basic process. The effect of heat pumping is only assessed *after* the process Grand Composite Curve has been established (see Section 2.3 of the Guide).

In the dryer area, the six drying loads have been combined into two to simplify the analysis. We see later that this has no effect on the result of the analysis. Also, the heating duties in the two zones are assumed to take place at the highest air temperature leaving each zone. This is because the small air temperature rise which occurs across each heater is unlikely to be exploitable by use of a counter-current heat exchanger. The load on the regenerator is the yearly average load. When it comes to the revamp design, we must remember that the regenerator load can rise to as much as 317 kW on hot and humid days.

Calculating the Problem Table (with a ΔT_{min} of 14°C) and plotting the results as a Grand Composite Curve yields the graph shown in Figure 4.14. The net hot utility requirement is 1414 kW, *without* any heat pumping. This compares with a net heat requirement for the existing design in Figure 4.13 of 1563 kW (remembering that heat in the utility steam condensate as well as the latent heat is utilised). If heat pumping is added to the profile shown in Figure 4.14 then significant energy savings must be possible. Note that the pinch is caused by the entry of the coldest dryer air stream into the problem, *i.e.* Dryer Zone 1.

4.4.3. Heat pumping strategy
The first point to note about the Grand Composite Curve shown in Figure 4.14 is that it is dominated by the large latent and "pseudo"-latent heat changes in the system. Another point to note is the fact that the only directly usable latent heat source is Flash Vapour 1. Flash Vapour 2 occurs below the pinch. Hence, to save more energy, heat from Flash Vapour 2 must be "lifted" above the pinch.

However, there is a difficulty. As shown in Figure 4.15 only 155 kW can be pumped across the pinch at minimum temperature lift. Further saving can only result from

Table 4.6

Stream type and Name	T(°C)	H(kW)	CP (kW/°C)	
Flash vap. 1 (hot)	79.4	1039	latent	
	79.4	0.0	heat	
Flash vap. 2 (hot)	43.3	974	latent	
	43.3	0.0	heat	
Condensate 1 (hot)	43.3	57	1.714	
	10.0	0.0		
Condensate 2 (hot)	86.9	232	3.017	EVAPORATOR AREA
	10.0	0.0		
Water (cold)	10	0.0	3.05	
	70	183		
Feed (cold)	37.8	0.0	3.98	
	87.8	198		
Flash 1 (cold)	79.4	0.0	latent	
	79.4	1005.5	heat	
Flash 2 (cold)	48.8	0.0	latent	
	48.8	643	heat	
Flash 3 (cold)	93.3	0.0	latent	
	93.3	173	heat	
Vent 1 (hot)	60	140	2.977	
	13	0.0		
Vent 2 (hot)	55	189	4.500	
	13	0.0		
Vent 3 (hot)	41	149	5.307	DRYER AREA
	13	0.0		
Regen (cold)	55	0.0	"pseudo" latent	
	55	93.4	heat	
Zone 1 (cold)	41	0.0	"pseudo" latent	
	41	254	heat	
Zone 2 (cold)	60	0.0	"pseudo" latent	
	60	124	heat	

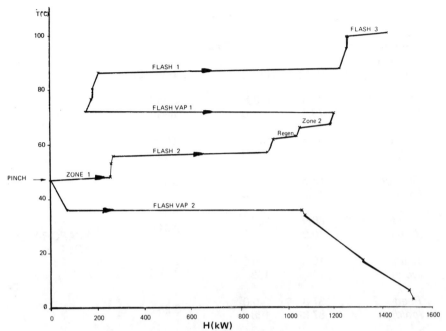

Figure 4.14—Base case Grand Composite Curve

heat pumped to a level hot enough to supply Flash 1. This represents a temperature rise of 50°C (allowing a $\Delta T_{\min}$ of 14°C) which is rather high for a heat pump. The situation can be improved by adopting a *two-stage* heat pumping strategy. Heat from the Flash Vapour 1 source is pumped to supply Flash 1, leaving a "hole" for hot utility supply at a lower level. Thus if a quantity of heat AB is pumped out of Flash Vapour 1, then the scope for heat pumping from Flash Vapour 2 at minimum temperature lift (*i.e.* into Zone 1 and Flash 2) becomes 155 kW + AB. We now note that heat pumping from Flash Vapour 1 up to supply Flash 1 *is already carried out in the existing design* by means of a steam ejector. It should be noted though that in the existing plant the steam ejector only takes 38% of the heat available in Flash Vapour 1 (*i.e.* AB out of AC) due to the fact that the "driver" high pressure steam mixes with the pumped vapour. If a mechanical compressor is adopted, then the potential for energy-saving becomes greater because maximising the heat which can be pumped from Flash Vapour 1 also maximises the heat which can be pumped from Flash Vapour 2 at minimum temperature lift.

Whatever means is adopted for pumping heat from Flash Vapour 1 into Flash 1, we must decide on a method for pumping heat from Flash Vapour 2. Clearly we could adopt compression by steam ejector or rotary mechanical device. However, a much neater solution presents itself. Namely, *we could lift the pressure of the whole evaporator train*. This would have the effect of raising the temperature of all the evaporator sources and sinks in "unison", but leaving the Dryer and Regenerator heat sinks in the same position. Thus the task of pumping heat from Flash Vapour 1 into Flash 1 becomes no more difficult, but Flash Vapour 2 becomes hot enough to drive the Dryer

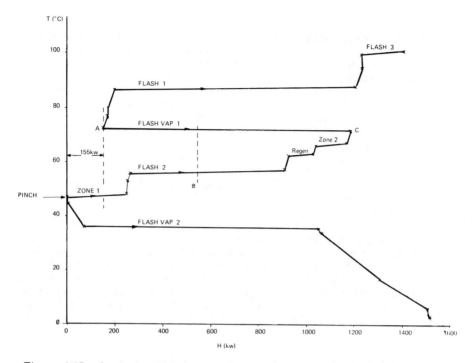

Figure 4.15—Analysis of base case Grand Composite Curve

and Regenerator without explicit use of heat pumps. Only if we require heat to be pumped from Flash Vapour 2 into Flash 2 will an actual heat pumping device be required.

Based on the above analysis we can identify two basic heat pumping schemes:

(i)   Raise the pressure of the evaporator train to allow Flash Vapour 2 to be used for driving the Dryer Zones and Regenerator, but leave the heat pumping scheme around the First Evaporator stage essentially as it is in the existing plant (but simply shifted upwards in pressure).

(ii) Raise the evaporator train pressure as in (i), but use a mechanical compressor to maximise the heat pumped from Flash Vapour 1, and then use a second mechanical compressor to pump heat available in Flash Vapour 2 into Flash 2.

Note that the extra capital cost involved in installing mechanical heat pumping equipment in scheme (ii) must be justified on the basis of the *marginal* savings available over scheme (i).

Referring back to Figure 4.13, all of the evaporator equipment is under vacuum. If the pressure of the whole train is to be raised without requiring hardware changes to the major capital items, then 1 bar is the pressure limit on Stage 1. In other words, the temperature in Stage 1 can be shifted to a maximum of 100°C and hence Stage 2 to a maximum of 86°C (0.6 bar), allowing for a 14°C drop. This maximum shift therefore allows a ΔT_{min} of 26°C between Flash Vapour 2 and the hot end of the Dryer Zones.

One final point can be noted about the design around Stage 3. Looking at the design we see that really, Stage 3 is not part of the pressure cascaded, or "multiple effect" system. It operates independently, directly on utility steam. Therefore if the pressure of the other two stages is raised, Flash Vapour 1 (at 100°C) will become hot enough to drive the Stage 3 Flash Heater (outlet temperature on the process side of the Stage 3 Flash Heater is 93.3°C). In fact, driving forces could be improved by increasing the recirculation flow around Stage 3 (the recirculation is present for outlet concentration control purposes). All of this reasoning leads us to ask why Stage 3 was not simply included as part of the double-effect system in the original design. The live steam injection which occurs between Stages 2 and 3 serves sterilisation purposes, but there is no particular reason why it must be located at this point. No satisfactory answer to the question could be found! There is no reason why Stage 3 should not be run on lower temperature steam, and the Flash 2 and Flash 3 loads can be included together in the double-effect system.

4.4.4. Evaluation of heat pumping schemes

(i) Heat pumping by steam ejector—Assuming that the heat pumping from Stage 1 Flash Vapour to the Stage 1 Flash duty is done by steam ejector, we obtain the mass balance shown in Figure 4.16(a). The total evaporation across Stages 1 and 2 has been held fixed at 2685 kg/hr (compare Figure 4.13) so as not to change conditions in Stage 3. This was thought desirable because of the temperature sensitivity of the product in concentrated solutions. The quantity of 0.6 bar steam available from Stage 2 for Dryer and Regenerator duty is 945 kg/hr, which is 205 kg/hr in excess of requirement (equivalent to 130.6 kW of "spare" heat at 86°C). However, this is only at the *average* yearly condition. On the hottest, most humid summer days, this surplus becomes a 186.4 kW deficit, which would then have to be made up from utility steam supply let down to 0.6 bar.

The scheme in Figure 4.16(a) only takes care of the latent (and "pseudo" latent) heat changes, based on the heat pumping analysis described previously. We must now check that the sensible heat sources and sinks are optimally interchanged. The stream data for this "residual" stream set are given in the diagram in Figure 4.17(a), along with the MER design that achieves the target predicted for this set. Notice that we have included the 130.6 kW of "spare" 0.6 bar steam from Stage 2, and the stream of process condensate ("Condensate 3") which is available at 86°C (previously, this was available as part of "Condensate 1" at 43.4°C). In the MER design, we have deliberately avoided matches to the hot Dryer Vent streams since such matches would be expensive and difficult to implement. The consequence of this decision is a small ΔT_{min} violation at the cold end of Match 4. Match 2 is insignificantly small, and so a sensible evolved version of the MER design is shown in Figure 4.17(b), with Match 2 eliminated and Match 4 merged into Match 1. This causes the ΔT_{min} violation to become worse. However a simple and acceptable structure has resulted. Match 1 already exists in the plant (see Figure 4.13), and is simply required to carry a higher load (211 kW rather than 130 kW). Since all exchangers in the existing system are of the plate-frame type, the increase in size can be implemented cheaply and easily by adding further plates. Since the other two matches are both to the cold water stream, they can be implemented in a single new frame. The residual utility steam heating duty on the feed stream could be incorporated into the Stage 1 Flash Heater, thereby saving

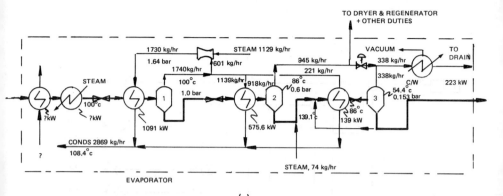

(a)

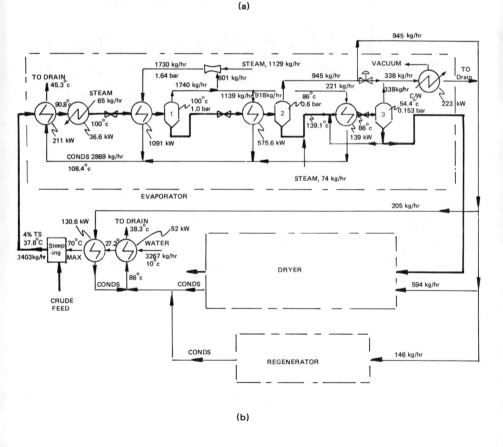

(b)

Figure 4.16—Evolved energy-saving scheme

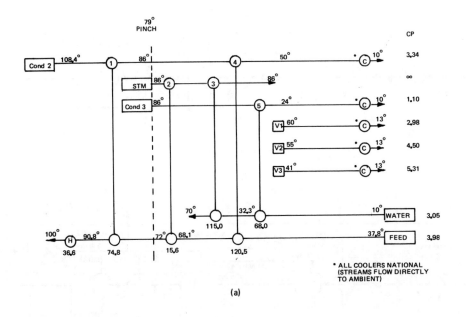

(a)

* ALL COOLERS NATIONAL
(STREAMS FLOW DIRECTLY
TO AMBIENT)

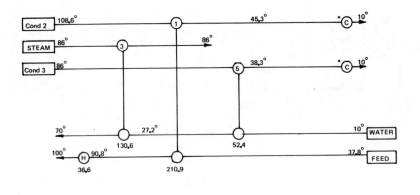

(b)

Figure 4.17— MER network design for sensible heat changes

222

about 13 kW of utility steam (due to the heat pump). This is a small saving, and it is probably better to leave some load on the Feed Heater for operability reasons.

The full revamp scheme is shown in Figure 4.16(b). The steam savings are as follows:

813 +	307 +	51 +	317 +	49 −	(1129 − 937)
(Dryer/Regen.)	(Stage 3)	(Injection)	(Water Heating)	(Feed Heater)	(extra to Ejector)

$$= 1345 \text{ kg/hr}$$

This improvement is worth £84 900 per year, representing about 50% of the plant steam demand. The effect of the heat pumping/pressure shift on the process source/sink profile is shown in Figure 4.18. Flash 2, Flash 3, and Flash Vapour 1 do not appear because they are exchanging heat at ΔT_{min}. The hot utility requirement is 785 kW, which compares with 1563 kW required in the base case, a 50% reduction. In other words, the revamp scheme achieves the target.

At the time of writing, preliminary costing has been completed, indicating a total installed cost for all the modifications of about £100 000. Around half of this cost is

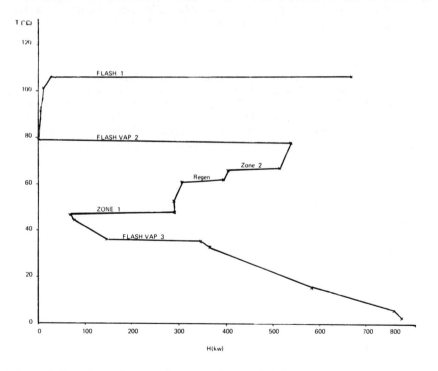

Figure 4.18—Grand Composite Curve for modified system

necessary for modifications to the Dryer. Under present operation, the mid-point ΔT between steam and Dryer air is 121 °C, which becomes squeezed to 45 °C in the revamp. Hence, a large amount of extra surface area is required, which in turn necessitates modifications to the drying tunnel and fans. An outline list of other modifications required is as follows:

(a) Change steam ejector.

(b) Control valve/pressure controller needed on the 0.6 bar steam system.

(c) Extra heat exchange area needed on Stage 3 Flash Heater and Feed Heater.

(d) Capacity of vacuum system needs increasing (due to increase in size of piping system under vacuum).

(e) Install 0.6 bar condensate system.

(f) Install new heat exchangers for cold water heating.

(g) Install let-down facility from utility steam system to 0.6 bar system.

With a Capital Grant available, the scheme achieves a payback of about 12 months.

(ii) Heat pumping by mechanical compressor—From Figure 4.18 it can be seen that the only *marginal* scope remaining for energy-saving by mechanical heat pumping is about 100 kW. Since this is of the same order of magnitude as the electricity which would be required by the heat pumps, the mechanical heat pumping option can definitely be discounted as uneconomic!

4.4.5. Conclusions

The main conclusion to be drawn from this study is that the Heat and Power synthesis concepts and techniques described in Section 2.4 of the Guide are not just for big plants in the bulk chemicals industry. They can yield surprising results on applications of much smaller scale. Also, they do not require the use of "black box" computer programs for finding good solutions. Rather, they lead the user into a thorough understanding of his problem by the systematic application of thermodynamic principles.

5. How to Apply the Principles in the Guide

5.1. General Strategy: Working with the Techniques

5.1.1.   Predicting minimum energy consumption

5.1.2.   Designing heat exchanger networks

5.1.3.   Changing the process to save further energy

5.1.4.   State-of-art improvements

5.1.5.   The overall retrofit strategy

5.2.   Overall Campaign: Monitoring, Targeting, Planning

5.1.   General Strategy: Working with the Techniques

If best results are to be obtained it is important that the thermodynamically-based techniques described in this Guide should be applied as part of an overall systematic methodology which also embraces state-of-the-art energy saving procedures and general process engineering skills (Figure 5.1). Furthermore there are several quite distinct stages in developing the full range of possible and realistic schemes:

5.1.1.   Predicting minimum energy consumption
The starting point is the gathering of reliable, meaningful data to describe current operations on the plant. This information will usually be obtained by taking on-plant measurements, using whatever instrumentation is available. Supplementary data may be needed to provide a comprehensive picture and this is usually gleaned from the plant manager and plant technical support team. The plant manual must be regarded as a last resort as there is often a considerable gap between documentation and current practice, particularly on a plant which has been operating for a few years!

The sequence of events which follows data collection is indicated in Figure 5.2. The data are first checked for credibility and consistency by carrying out the appropriate process engineering calculations. For example, mass and heat flows across the plant must be seen to balance; phase calculations must confirm that the known conditions—for instance, at the base of a distillation column—are consistent with temperature and pressure; estimated heat loads on exchangers must correspond to thermal rating checks on the unit; *etc.*

IDENTIFY RETROFITS BY COMBINING:-

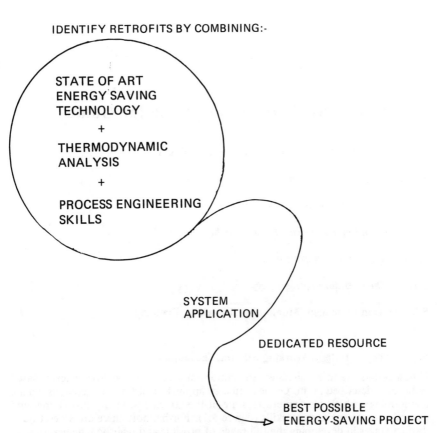

Figure 5.1—An overall systematic methodology should embrace thermodynamically-based techniques, state-of-the-art energy saving procedures and general process engineering skills

Once the data are accepted as reliable it is a relatively simple matter to run the minimum energy consumption algorithm and then compare the predicted minimum utilities requirement with the actual on-plant consumptions. Typical figures for a modern naphtha reformer plant investigated in ICI are shown in Table 5.1.

Table 5.1—Scope for Further Integration of 1300 Unit Aromatics II

System	Saving £m	% minimum Achievable Heat Load
Current Operation	—	123.1
Minimum	2.1	100.0

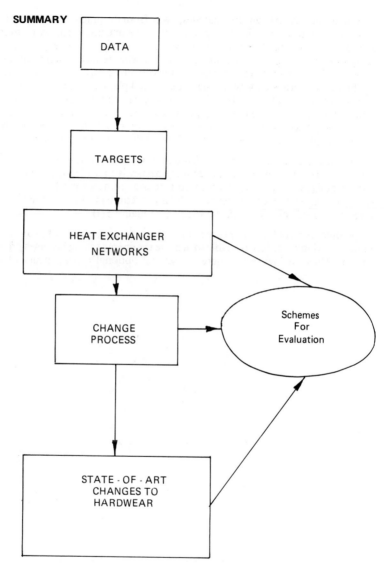

DATA

TARGETS

HEAT EXCHANGER
NETWORKS

CHANGE
PROCESS

Schemes
For
Evaluation

STATE - OF - ART
CHANGES TO
HARDWEAR

Figure 5.2—Checking data for credibility and consistency

5.1.2. *Designing heat exchanger networks*

Once the minimum energy usage has been calculated for the existing plant arrangement, a target exists towards which the design engineer can work.

The maximum energy recovery design is now arrived at by designing a workable system at, and around, the critical pinch region. Once this part of the system is defined, the remaining heat exchangers' matches, away from the pinch, are comparatively easy to achieve. In practice it is usually found that the maximum energy recovery design does not represent the best trade-off between capital cost and energy saving. A much faster pay-back is likely to be achieved if a somewhat reduced energy recovery can be tolerated. Furthermore, in carrying out retrofit projects, it is essential that normal production patterns on an established operational plant are unaffected; if a retrofit is proposed which shuts the plant down for a greater time than the routine planned maintenance period, then the subsequent loss of revenue will have to be shouldered as part of the project costs. Therefore, project work should where possible, be phased into short sub-projects which interfere little with the normal running of the plant.

A retrofit strategy is required which attempts to achieve this production-sympathetic approach. The maximum energy recovery design is first identified and then several further designs are produced which correspond to increasing levels of relaxation, away from maximum energy recovery. This procedure gives rise to the evolution of a series of phased complementary projects (see Figure 5.3).

It should be noted that in proceeding with succeeding phases of project work, the number of viable options available to achieve the energy saving becomes progressively smaller. Thus, in Figure 5.3 there are several optional arrangements which could

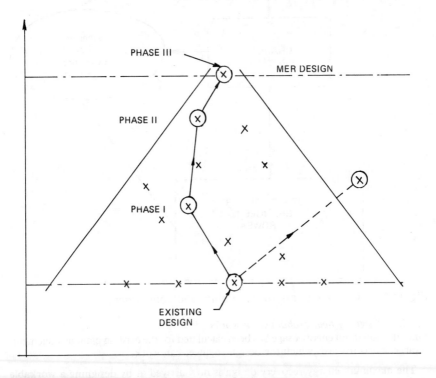

Figure 5.3—Effect of unsystematic retrofit modification

Table 5.2—Achievement on Aromatics II, 1300 Unit

System	Saving	%Minimum Achievable Heat Load
Current Operation		123.1
	Δ = £0.5m	
Phase II		117.9
	Δ = £1.0m	
Phase III		106.6
Minimum		100.0

achieve the energy recovery level of the existing design (each denoted by a cross). However, there are progressively fewer crosses (or options) available at Phase I and Phase II level.

It can be seen that a funnel effect exists in progressing towards maximum energy recovery. If this effect is not recognised and a Phase I "option" is chosen which falls outside the funnel (Figure 5.3) then subsequent phases of energy recovery will not be possible and much of the available potential for saving will be lost. In other words, the systematic approach is essential and haphazard uncoordinated energy-saving—with no regard for the total system—should definitely be avoided.

Table 5.2 illustrates how three phased projects were developed to approach the target energy saving on a naphtha reformer. The economic trade-off point was found to be 106.6% of minimum achievable heat load in order to ensure highly attractive projects with pay-backs of the order of six to seven months.

The corresponding changes to the plant arrangement are discussed in the worked example, Section 4.3.

5.1.3. Changing the process to save further energy

It has already been mentioned that a systematic thermodynamic analysis reveals the minimum energy consumption which is possible for a fixed plant arrangement as specified by the input data. If the processing arrangements are suitably modified, it is sometimes possible to reduce this minimum energy consumption.

In carrying out the initial target prediction for the existing plant arrangement and the subsequent heat exchanger network study the attention of the engineer is concentrated at the critical points in the system where he must achieve a design in which heat flows and heat integration are particularly well-balanced. It is inherent in the systematic approach that the engineer will be aware of such factors as the pinch point, and often he will feel constrained by fixed stream flows and temperatures which limit his ability to integrate the system. Thus, he is stimulated into lateral thinking at such critical areas and often he will evolve possible modifications to the processing route.

In effect, such modifications imply a fundamental change in the input data and it is imperative to return the problem through the targeting and networking stages in order to verify that, in terms of total system, the process change really does reduce the minimum energy consumption.

The systematic procedure described here is a great help in identifying the important areas in the plant arrangement where modifications to the processing route could save energy. However, it is of course conceivable that some of the possible modifications could be spotted by an experienced engineer working in a more conventional process engineering mode. In fact, some may be so obvious that he would have difficulty in not identifying them!

However, haphazard selection of energy recovery devices may not always be appropriate when viewed in terms of the total system. Thus systematic examination of a large tonnage plastic monomer plant revealed that the part of the processing system, which was designed to incorporate 3 MW of heat recovery was inappropriately placed across the pinch and, in terms of the total system, was ineffective. Modifications were subsequently carried out to use the waste heat at a more appropriate place in the process.

It is apparent, then, that the overall effect of a process change on the total system will only be revealed through a systematic analysis and indeed, this cross-check is essential before proceeding with such proposals. But, in addition to this, it is clear that the more ingenious, less obvious options—which ultimately provide the edge over the competition—are most likely to be brought to light by the systematic approach. Thus, heat recovery studies on the desulphurisation stage of a naphtha reformer have provided new perspectives concerning the feed/effluent heat interchange which is a standard feature of such units. Modifications, which are currently being patented, have been defined and will save £0.6 M *p.a.* when installed on the ICI unit.

5.1.4. State-of-art improvements

It is interesting to observe because of long implementation times that almost all large-scale process plants currently in operation were built from designs which had little emphasis on energy saving. It is only some eight years after the original oil crisis that a strong momentum has built up in the general area of energy saving technology. In terms of plant hardware, this means that systems and equipment are now emerging which operate to considerably higher energy efficiencies and levels of heat recovery than those specified as original equipment a few years ago.

In carrying out a comprehensive energy revamp of an existing installation, it is therefore imperative that maximum use is made of general improvements in the state-of-art hardware, in addition to pursuing the areas already described here. Usually, it is desirable to arrive at a complementary blend of retrofit schemes emerging from all areas.

There are many areas in which major improvements have taken place in recent years but of these waste heat recovery must deserve specific mention. With the proprietary equipment now available it is usually economic to retrofit for recovery of heat from stacks down to dew point levels (170°C).

The operation of furnaces and boilers can also be improved dramatically by appropriate modifications. Control of excess air, retilling for improved flame-shape and the use of on-line control are all effective, short pay-back options.

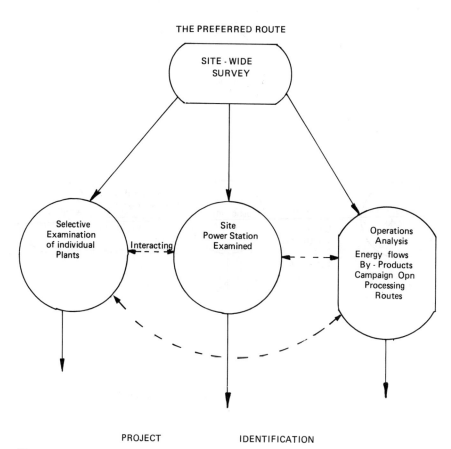

THE PREFERRED ROUTE

SITE - WIDE
SURVEY

Selective
Examination
of individual
Plants

Interacting

Site
Power Station
Examined

Operations
Analysis
Energy flows
By - Products
Campaign Opn
Processing
Routes

PROJECT   IDENTIFICATION

Figure 5.4—Preferred route for a retrofit strategy

The general area of computer control has taken major steps forward with the advent of the silicon chip. General plant control by micro-computer and operator feed-back by data logger offer important avenues for saving energy.

Developments in the operating temperatures of gas turbines have improved their efficiency levels dramatically and they are now becoming serious contenders for prime power supply in either on-plant or combined heat and power station situations. Similarly, turbo-expanders now offer a viable means of power recovery from process off-gas and waste steam. The general area is exciting and there is far more to come.

5.1.5.   The overall retrofit strategy
It has already been mentioned that a general approach to retrofit projects has evolved which is summarised in Figure 5.2. The design stages described in 5.1.1 to 5.1.4 come together into an interactive procedure. In particular, considerable feedback can occur between changes to the process and the earlier stages of targeting and networking.

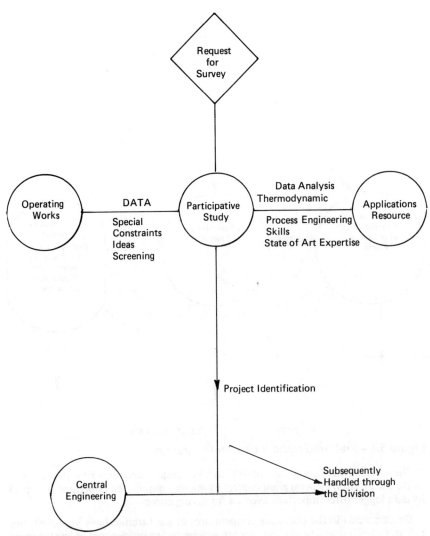

Figure 5.5—How the system works

The preferred scope for a study should be site-wide. Such a broadly based survey ensures that no stone is left unturned. Besides carrying out systematic examinations of individual plants and the site power station (if there is one), it is then possible to examine the implications of interactions between plants and interaction of plant with power station.

Also, aspects of the general site operations can be investigated with a view to bringing about improvements in such areas as use of by-products as fuel, optimising overall processing routes and planning energy-saving campaigns (Figure 5.4).

The final aspect of strategy—and probably the most important—is to ensure that a spirit of co-operation is established with operating staff. Ideally, the investigation should be regarded as a participative venture with the works' staff. In this way the best blend of projects is achieved: the works personnel will provide basic heat and mass balance data and indicate where there are constraints in the scope for redesign. After the design engineer has applied his techniques and identified a preliminary selection of potential projects, the works' man will be invaluable in the subsequent identification of impracticalities and in the superimposing of further ideas. The general mode of working is shown in Figure 5.5.

5.2. Overall Campaign: Monitoring, Targeting, Planning

Without doubt, it is preferable to produce a reasonably detailed plan which extends over a period of three to four years. In this way the relative importance of the various elements can be brought into perspective with respect to time and it is then possible to plan short-term activities in greater detail.

Before setting up the plan, it is worthwhile considering what would happen if no attempt was made to save energy. Investigations suggest that the contribution of

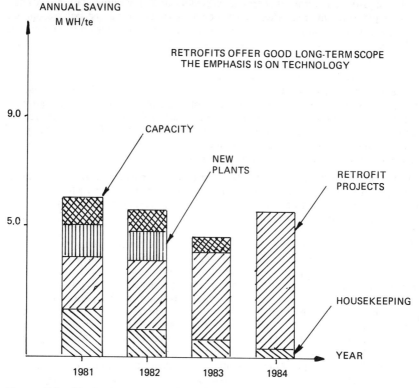

Figure 5.6—Typical energy-saving plan

233

energy costs to the selling price of chemicals has already risen significantly and will continue to rise. Furthermore, in order to peg the effect of rising energy prices, it is considered that an annual reduction in usage of approximately 5% (in terms of MWH/tonne) is essential.

Figure 5.6 represents a typical four-year plan for a chemical processing complex where the declared objective is to reduce specific energy consumption by 5% *p.a.* In this particular works, individual manufacturing plants are currently (1981) running at less than full capacity and the first element in the plan refers to the expected work-up to full capacity over the next two years; as this takes place, energy efficiencies will increase and there will be an overall reduction in specific consumption.

A second element in the plan refers to an ongoing campaign which is dedicated to energy saving *via* housekeeping measures. With the available allocation of maintenance and plant resources, it is envisaged that the law of diminishing returns will apply and that the housekeeping element will become progressively less effective in achieving further savings, over the period of the next few years.

A third and important element is the provision for energy improvements by bringing on stream new, high-efficiency plants. Obviously the introduction of new plants cannot be justified in terms of energy savings alone and only a limited number of new units are planned—in areas where market demands make this a justifiable economic investment. In the example taken, a new plant is expected to come on stream in 1982 and make its presence fully felt, by reduced specific energy consumption, over a period of 24 months.

The elements discussed so far do not add up to the target reduction of 5% in any one year of the four-year plan. It is a simple matter to deduce, therefore, that the shortfall must be made up by retrofit energy projects and it can be seen in Figure 5.5 that in a capital constrained situation, where the scope for new plants is limited, the mainstay in continued energy saving lies with retrofit projects. In fact, as capacity and housekeeping effects decline in importance over the four-year period, the role of retrofits becomes progressively more important.

It can be argued that this is only the case for the particular set of circumstances chosen in the example. However, the conditions selected are those which are generally prevailing in major continental Western Europe and USA installations and as such the general conclusion is broadly tenable.

An overall plan which places increasing reliance on retrofits has important implications concerning ongoing allocation and planning of capital resources. Experience in ICI indicates that during the initial period of retrofit project work (say the first two years) there will be an adequate number of projects identified with pay-back of one year or less. This gives a pointer as to capital allocation for this period; *i.e.* capital allocation should approximately equal projected annual energy saving (in money terms). In succeeding years, it can be expected that paybacks will become longer as the simpler, fast pay-back projects are preferentially selected first. Proportionally greater sums will therefore have to be allocated for later periods in order to maintain the initiative. The realistic planning of capital requirements is one of the most important aspects of prolonged energy-saving activities.

An equally important aspect of planning is to anticipate requirements in terms of human resources. It is apparent from earlier sections of this paper that effective energy saving is heavily dependent on highly trained, specialised, technical resources. In terms of retrofit project work, there is a limit to the numbers of projects which an experienced, trained engineer can handle in one year. Great care must be taken to ensure that adequate engineers are available and it is noteworthy that as time goes by and pay-back times inevitably lengthen (*i.e.* energy saved per project becomes less) then it will be necessary to have an ever increasing number of projects under way. The implication is that greater numbers of engineers will have to be employed to maintain the annual reduction.

It is important to observe that meaningful medium-term planning is dependent on the existence of a reliable system for monitoring on-plant energy consumption. The preferred basis for such a system is MWH consumed per tonne of product made. It is essential that corrections are incorporated for capacity-related effects (efficiency varies with output) and product mix effect (MWH/tonne changes with variation in product spectrum). Ideally, the system should be based on existing data sources and should be computerised.

Ultimately, there is a limit to how much energy can be saved and after a period of years it will be necessary to re-examine the declared annual target for saving and decide whether a relaxation is necessary. From experience in ICI, it would appear that there is a considerable way to go before this becomes necessary.

6. Concluding Remarks

6.1. University Teaching

6.2. Technology Transfer

6.3. Where Do We Go from Here?

The design of most large industrial processes is based on a long period of development with many evolutions and improvements leading to today's flowsheet. Naturally we therefore tend to assume that these flowsheets are more or less optimal, with no big "faults" left in them. What a surprise then to discover, say, that there is room for major improvement in "state-of-the-art" flowsheets and that in many case studies carried out using the techniques described in this Guide (see Table 1.1 in the Introduction) energy savings could be secured *alongside* capital savings!

All this seems to be due to one major discovery: the heat transfer "Pinch". Without knowing about the existence of the "Pinch", the implications of heat transfer across it and of heat engine placement above or across it, *etc*, how can the designer of a large and complex integrated plant get everything right? Following the discovery of the "Pinch", we expect that a quantum change in the performance of integrated chemical plant will take place: the average energy savings identified in the case studies listed in Table 1.1 were 30%!

How then do we tackle the task of technology transfer to make this quantum change come about? What is the best way of teaching new technology in universities? And, last but not least, how do we develop the technology further to take advantage of the possibility of "downstream" discoveries? These issues are now addressed one by one.

6.1. University Teaching

Process design and network integration techniques have been included, if only in preliminary form, in a number of university chemical engineering courses in the last ten years. Often the material has been tacked on to existing schemes, creating artificial marriages with economics and design on the one hand while remaining divorced from thermodynamics and heat transfer technology on the other. However even BSc syllabi change and there are many opportunities now emerging for the inclusion of new material in appropriate places in the enhanced courses of the 1980s.

237

One factor often forgotten is that students learn by a stagewise process. Consequently it is necessary to introduce the concepts and techniques at different points in the course.

There is no doubt that the heat cascade principle and the pinch concept can and should be introduced at the first year level and should relate to thermodynamics, unit operations and heat transfer.

Networks on the other hand require a more mature outlook on design and create problems if introduced too early. Much of the power of the new methods described here lies in the implicitly assumed broad background of the user. Much of the appeal similarly lies in the newfound ability to organise much of the intuitive hunch of the experienced designer. Students will not have the experience to recognise these points.

Finally the case studies will form an invaluable source of material in design teaching at the end of the course. Chemical engineers have always recognised the value of the design project to integrate the various parts of the syllabus, enabling the student to digest his course. So these case studies follow this old theme using the novel concepts and techniques.

6.2. Technology Transfer

There are three means of technology transfer one might think about for introducing the techniques covered in this Guide into everyday design practice: general education, specialist computer software and specialist experts.

General education
As indicated above, the techniques would seem important enough to be included in general undergraduate university education. At the same time, application of the techniques in earnest requires a great deal of familiarity and less than regular use would leave the general user "out of practice". Thus, general education will, in the long run, provide a situation where most chemical engineers ought to be able to identify potential applications, but before long will require specialist software or will have to call in a specialist expert.

Specialist software
Computer programs for energy networking are available and some are based on Linnhoff's earlier techniques (Linnhoff and Flower, 1978). However, these earlier techniques did not recognise the criteria for correct stream splitting and did not include facilities for constraints. Also, all programs known to the authors to date are non-interactive.

This last feature means that it is necessary, in order to run the software, to isolate the energy networking problem from the overall process design. It is then up to the user to identify any advantage that would be gained from changing both process and energy network at the interface. The outline case studies in Section 4 give many examples where the difference between success and failure in a project was due to the ability to deal with this interface properly.

The lack of a constraint facility (*i.e.* the software assumes that any hot stream can match with any cold stream) combined with the fact that the user cannot "interact" with the program, means that the software is difficult to use in the context of retrofit work. The input data have to be patiently manipulated to generate a network on the "unconstrained" computer that will resemble an existing process to such an extent that the differences amount to a practical retrofit project. And even then, there is no guarantee that this retrofit project is the most rewarding one.

Having said all this, the idea of targets is incorporated in some of the programs and for the engineer who knows little about the principles covered in this Guide, this facility on its own must make it worthwhile to use the programs.

Summarising, the present "state-of-the-art" in energy network software does not yet convince, but is certainly better than nothing. Hopefully, the packages will soon be brought up to date (stream splitting, constraints, etc) and be made interactive as software does seem to be a promising and sensible way forward for widespread implementation of the new technology.

Specialist experts
A specialist expert in general design techniques is not a narrow expert, but someone who must be capable of applying his expert knowledge over a wide range of processes, technologies and objectives. He has to link successfully the network integration technology with the specific technology of whatever process he is examining and has to identify those opportunities for savings at the interface between the process and the energy network that are so essential for successful application of the new methods. Given that software cannot identify these interface opportunities, experts seem to be the most promising route forward for fast impact. The training of specialist experts is indeed the route chosen for fast technology transfer by ICI and other companies after considerable evaluation.

6.3. Where Do We Go From Here?

At the end of the Guide, let us try and see the subject in perspective. How much else is there to know in heat exchanger networks and in heat and power integration? And how much else is there to know in process design?

On the basis of the material collected in this Guide many might say that heat integration and heat and power integration are now mature areas. However, there is a whole host of questions yet to be resolved:

(1) How do we properly exploit process flexibilities (such as reaction temperatures, distillation column pressure and reflux ratio, side stream take-off points, etc)?

(2) How do we design for good operability?

(3) How do we allow systematically in our designs for constraints such as forbidden and imposed matches?

(4) How can we target more precisely for capital cost than by simply predicting the "number of units"? How can we take into account individual stream film heat transfer coefficients, materials of construction, engine size and efficiency, etc?

(5) How can we deal properly with the problem of trade-offs in complex design? How can we guess the best value for ΔT_{min}? How can we trade-off the number of units against stream splits against total surface area against energy?

(6) How can we extend the techniques so as to give us good systematic procedures away from the Pinch?

(7) How do we modify the techniques so that they become explicitly applicable to "retrofit" designs?

From this list, the reader will realise that the material presented in this Guide, while being perfectly usable in its own right, only represents the beginning of what will soon be a body of technology of considerable depth.

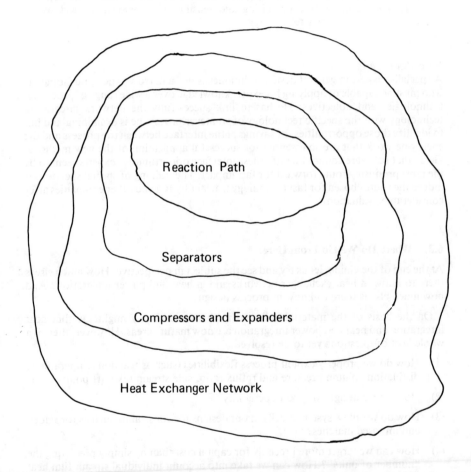

Figure 6.1—The hierarchical nature of a chemical process design

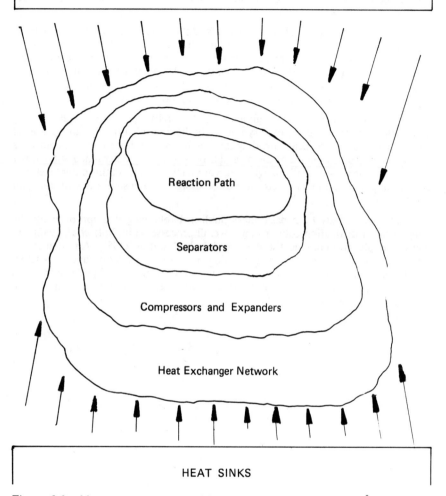

Figure 6.2—Most processes are ultimately driven by heat cascading through them

As far as the design of overall processes is concerned, consider Figure 6.1. The sketch shows symbolically the hierarchical nature of a chemical process design: to begin with, the designer has to know his chemical reaction "path", *i.e.* the feeds, the products and the recycles involved. Then, and only then, can he determine the separation tasks. After this, he can gain a feel for pressures, temperatures and phase equilibria and can determine the need for pumps, compressors and expanders. Finally, now that the various heating and cooling duties (for, say, reactor feeds, distillation reboilers, compressor intercoolers, *etc*) are known, he can tackle the heat integration task. In other words, there is a hierarchical build-up and there would be little point in trying to design the heat exchanger network as long as the separators are not yet agreed upon. The design develops from the "inner level" of Figure 6.1 to the "outer level".

If this observation seems obvious, then it is all the more interesting to note that the approach to heat and power integration as introduced in Section 2.3 follows the opposite approach: the heat engine design is selected by working from the outermost level (heat exchanger network) to the next inner level (heat and power). In essence, we examine the shape of the composite curve of the heat recovery problem and then select a heat engine so as to modify the shape of the composite curve!

Although it contradicts the argument of "hierarchy" put forward above, this approach seems to make sense: as symbolically shown in Figure 6.2, most processes are ultimately driven by heat cascading through them from hot sources to cold sinks and the energy performance of the overall process is largely determined at the outer level of the "process onion". So why not work backwards from this outer level thus determining how the inner levels have to be designed to ensure a low overall heat flow?

There have already been basic ideas published following this approach for the integration of distillation columns into overall processes (Linnhoff et al, 1982) and many examples from practical studies are given by Linnhoff (1982). An extension of these ideas into the treatment of separation other than distillation and of reactors clearly seems possible. In other words we can foresee future techniques helping us to design total processes with the same degree of confidence which we now have when dealing with simple heat integration.

References

Boland, D. and Linnhoff, B. (1979) "The Preliminary Design of Networks for Heat Exchange by Systematic Methods", *The Chemical Engineer*, April, 222.

Cerda, J., Westerberg, A.W, Mason, D.R. and Linnhoff, B. (1982) "Minimum Utility Usage in Heat Exchanger Network Synthesis—A Transportation Problem", to appear in *Chemical Engineering Science*.

Cohen, H., Rogers, G.F.C. and Saravanamutto, H.I.H. (1972) "Gas Turbine Theory", Longman, ISBN 0 582 44926 X.

Cooper, P.R. (1981) "Royal Society Esso Energy Award". Press Release 8(81), Royal Society (London).

Dunford, H.A. and Linnhoff, B. (1981) "Energy Savings by Appropriate Integration of Distillation Columns into Overall Processes", paper No. 10, Cost Savings in Distillation symposium, IChemE, Leeds.

Flower, J.R. and Linnhoff, B. (1979) "Thermodynamic Analysis in the Design of Process Networks", *Comp. and Chem. Eng.*, **3**, 283.

Haywood, R.W. (1980) "Analysis of Engineering Cycles", Pergamon (3rd Edition), ISBN 0 08 025441 1.

Hinchley, P. (1977) "Waste Heat Boilers: Problems and Solutions", *Chem. Eng. Prog.*, **73**, 90 and *Chem. Ing. Tech.*, **49**, 553.

Hohmann, E.C. (1971) "Optimum Networks for Heat Exchanger", PhD Thesis, University of S. California.

Linnhoff, B. (1979) "Thermodynamic Analysis in the Design of Process Networks", PhD Thesis, University of Leeds.

Linnhoff, B. (1982) "Design of Heat Exchanger Networks—A Short Course", Department of Chemical Engineering, UMIST, P.O. Box 88, Manchester.

Linnhoff, B. and Carpenter, K.J. (1981) "Energy Conservation by Exergy Analysis— The Quick and Simple Way", Second World Congress of Chemical Engineering, Montreal, Canada.

Linnhoff, B., Dunford, H. and Smith, R. (1982) "Heat Integration of Distillation Columns into Overall Processes", submitted for publication to *Chemical Engineering Science*.

Linnhoff, B. and Flower, J.R. (1978) "Synthesis of Heat Exchanger Networks" (2 parts) *AIChE Journal*, **24**, 633.

Linnhoff, B. and Hindmarsh, E. (1982) "The Pinch Design Method of Heat Exchanger Networks". Full version submitted to *Chemical Engineering Science* (1982); short version presented at Understanding Process Integration Symposium, IChemE, Lancaster.

Linnhoff, B., Mason, D.R. and Wardle, I. (1979) "Understanding Heat Exchanger Networks", *Comp. and Chem. Eng.*, **3**, 295.

Linnhoff, B. and Turner, J.A. (1980) "Simple Concepts in Process Synthesis Give Energy Savings and Elegant Designs", *The Chemical Engineer*, December, 742.

Linnhoff, B. and Turner, J.A. (1981) "Heat Recovery Networks: New Insights Yield Big Savings", *Chemical Engineering*, 2 November, 56.

Masso, A.H. and Rudd, D.F. (1969) "The Synthesis of System Designs, II. Heuristic Structuring", *AIChE Journal*, **15**, 10.

Smith, R.A. (1981) "Economic Velocity in Heat Exchangers", presented at the ASME/AIChE 20th National Heat Transfer Conference, Milwaukee.

Taborek, J. (1982) "Mean Temperature Difference". Chapter 1.5 of "Heat Exchanger Design Handbook" (Eds E.U. Schlunder, K.J. Bell, G.F. Hewitt, F.W. Schmidt, D.B. Spalding, J. Taborek and A. Zukauskas), Hemisphere Publishing Corporation, New York and Washington.

Townsend, D.W. and Linnhoff, B. (1982a) "Heat and Power Networks in Process Design" (2 parts), submitted to *AIChE Journal*.

Townsend, D.W. and Linnhoff, B. (1982b) "Designing Total Energy Systems by Systematic Methods", *The Chemical Engineer*, March, 91.

Index